T0138791

Cognitive Radio Networks Optimization with Spectrum Sensing Algorithms

RIVER PUBLISHERS SERIES IN COMMUNICATIONS
Volume 44

Series Editors

ABBAS JAMALIPOUR
The University of Sydney
Australia

MARINA RUGGIERI
University of Rome Tor Vergata
Italy

HOMAYOUN NIKOOKAR
Delft University of Technology
The Netherlands

The "River Publishers Series in Communications" is a series of comprehensive academic and professional books which focus on communication and network systems. The series focuses on topics ranging from the theory and use of systems involving all terminals, computers, and information processors; wired and wireless networks; and network layouts, protocols, architectures, and implementations. Furthermore, developments toward new market demands in systems, products, and technologies such as personal communications services, multimedia systems, enterprise networks, and optical communications systems are also covered.

Books published in the series include research monographs, edited volumes, handbooks and textbooks. The books provide professionals, researchers, educators, and advanced students in the field with an invaluable insight into the latest research and developments.

Topics covered in the series include, but are by no means restricted to the following:

- Wireless Communications
- Networks
- Security
- Antennas & Propagation
- Microwaves
- Software Defined Radio

For a list of other books in this series, visit www.riverpublishers.com
http://riverpublishers.com/series.php?msg=Communications

Cognitive Radio Networks Optimization with Spectrum Sensing Algorithms

Tanuja Shendkar Dhope

G. H. Raisoni College of Engineering and Management,
Pune, India

River Publishers

Published, sold and distributed by:
River Publishers
Niels Jernes Vej 10
9220 Aalborg Ø
Denmark

River Publishers
Lange Geer 44
2611 PW Delft
The Netherlands

Tel.: +45369953197
www.riverpublishers.com

ISBN: 978-87-93102-00-2 (Hardback)
       978-87-93102-01-9 (Ebook)

©2015 River Publishers

Dedicated to my family
My beloved husband Satish and dearest son Pranav
My father Chandrakant Shendkar,
My father-in-Law Captain Krishna Dhope
My mother Lilawati Shendkar,
My mother-in-law Shalutai Dhope

Contents

Preface

The evolution of modern wireless communication systems and the upgrade of existing system demands for higher throughput and higher mobility, which are on the other hand challenged by radio scarcity in allocated electromagnetic spectrum bands. The existing fixed spectrum allocation scheme in wireless communication results in underutilization of spectrum, thus leaving open the possibility of negotiating with the existing systems and utilizing the unused part of the spectrum which is called "Spectrum holes/White Spaces" for communication. This concept of opportunistic utilization of unused part of the spectrum in wireless communication is the main idea for **Cognitive Radio**, which adapts the radio parameters based on knowledge of the environmental radio sources.

Around the globe, there is a coordinated move for "Digital Switchover" by discontinuing analogue TV transmission. TV switchover to full digital broadcast service created new spectrum opportunities due to higher spectrum efficiency compared to analogue services. This Digital Dividend consolidates "leftover" frequencies resulting from the change of TV broadcasting from analogue to digital. These cleared bands due to digital dividend and unused TV channels are called TVWS. The TVWS attracted many researchers due to its excellent propagation characteristics compared to GSM 1800 MHz, 2.1 GHz or 2.5 GHz bands, making broadband wireless access cheaper.

The survey of spectrum utilization for 600–800 MHz that is allocated to TV channels in India shows that spectrum in these bands is underutilized and can be opened for cognitive radio applications and other advanced new wireless services for next-generation networks because of its excellent signal propagation characteristic where radio signals better penetrate building walls and floors when the individual radios move outside the coverage footprint of the communication system.

In cognitive radio, sensing a spectrum is a crucial task. The purpose of this book was to optimize cognitive radio networks through spectrum sensing algorithms in the underutilized spectrum in wireless network and to improve throughput requirement of new high-speed technology.

The aim of this theme was to explore new techniques of spectrum sensing in cognitive radio networks utilizing TVWS created by digital dividend due to analogue transmission shutoff in wireless communications, and the possibility of its application for Indian scenario under different propagation models. The work also justifies the TVWS opportunities and regulatory aspects in India for cognitive radio applications and the use cases for exploitation of TVWS depending on user's mobility and geo-location between user and base station. It will contribute to analysis of cooperative power sensing of the received signal for spectrum sensing to detect whether the received signal is signal of interest or noise. The sensing will be based on the knowledge of noise power, analysing its performance under different channel models such as Rayleigh and Rician channels and comparing the performance of energy detection with covariance absolute value method. It further explores the application of DoA estimation algorithm for the desired signal in the cognitive radio networks.

This book is organized into four main chapters:

First chapter describes novel applications of TVWS. This chapter gives insight to the readers about the digital dividend activities at international level including band plan for 698–806 MHz band for IMT applications and elaborates the DD scenario in India. The proposal from Joint Task Group has been constituted to study the compatibility between services/applications in 700 MHz band in India, and the opportunistic spectrum access in India focusing on the candidate frequencies for CR technology has been studied in this chapter. Further, the various opportunities for exploiting TVWS in an efficient manner including IEEE 802.11af standards, broadcasting services, femto-cell for wireless broadband, public safety application and how efficiently this TVWS can be utilized for rural India, the key use cases and scenarios for TVWS opportunities and the regulatory activities related to CR and TVWS are given in this chapter.

Second chapter considers different aspects of dynamic spectrum access and overviews the various spectrum sensing challenges and study of different types of spectrum sensing algorithms together with advantages and disadvantages of each method. The analysis of energy detection without noise uncertainty and with noise uncertainty and covariance absolute value method-based DVB-T standard using MATLAB under different time-varying fading channels such as Rician and Rayleigh is evaluated. The need for hybrid detection algorithm is identified by comparing energy detection and covariance absolute value. The simulation results for hybrid detection method for detection of DVB-T standard under different time-varying fading channels such as Rician and Rayleigh have been studied.

Third chapter focuses on the need for cooperative sensing and classification of cooperative sensing based on how cooperating CR users share the sensing data in the network: centralized, distributed and relay-assisted. The simulation results based on cooperative power sensing of the received signal for spectrum sensing to detect whether the received signal is signal of interest or noise, and based on energy detection with hard combining for "logical AND" and "logical OR" are given.

Chapter four considers the performance of DoA estimation algorithms such as Bartlett's, Capon, MUSIC, ESPRIT and root-MUSIC based on number of snapshots, signal-to-noise ratio (SNR), MSE, standard deviation of DoA, and user space distribution such as narrow angular separation, wide angular separation and combination of narrow and wide angular separation. Further, performance of DoA estimation algorithm in CR context is analysed in time-varying fading channel. The new adaptive techniques have been proposed to improve the performance of DoA algorithms, and a novel, better performing algorithm has been proposed.

Acknowledgments

All praise, land and honour to "Lord Shree Ganesha" and "Maa Sarasawti".

I would like to express my gratitude to my supervisor, Prof. Dina Šimunić, whose expertise, guidance, and understanding helped me at every point during my research at University of Zagreb, Croatia.

I would like to give special thanks to Prof. Sandeep Inamdar and Prof. Ramjee Prasad, Aalborg University, Denmark for giving me an opportunity for doing my research work through 'Erasmus Mundus Mobility for Life Programme'. I take an opportunity to thank management and Principal, G. H. Raisoni College of Engineering and Management, Pune, Maharashtra, India for the encouragement.

I am persuaded to express gratitude my Husband Satish Dhope for his love, understanding and encouragement. I am grateful to my family: my mother Lilawati Shendkar, my father Chandrakant Shendkar, my mother-in-law Shalu Dhope, my father-in-law Capt. Krishna Dhope, my sister Anita Bhosale, my brother Manoj Shendkar, my brother-in-law Milind Bhosale, my sister-in-law Shital and Archana for their understanding and the support they provided me with through my entire life and career.

Thanks to all those whose names are not mentioned here, but were directly or indirectly associated with this task.

Tanuja Shendkar Dhope

List of Figures

List of Tables

List of Abbreviations

ARCEP	French Electronic Communications and Postal Sector Regulator
ASO	analogue switch off
BAS	broadcast ancillary services
BPS	broadcast programme making services
BWA	broadband wireless access
CAV	covariance absolute value
CLT	central limit theorem
CR	cognitive radio
DD	digital dividend
DoA	direction-of-arrival
DoT	Department of Telecommunication
DSA	dynamic spectrum access
DSO	digital switch off
DSP	digital signal processor
DTTV	digital terrestrial TV
DVB-H	digital video broadcast-for handheld
ED	energy detection
EIRP	effective radiated isotropic power
ETSI	European Telecommunication Standards
FCC	Federal Communications Commission
FPGA	field-programmable gate array
HD	hybrid detection
HDTV	high-definition TV
ITU	International Telecommunication Union
LRT	likelihood ratio test
MIMO	multiple-input multiple-output
NFAP	National Frequency Allocation Plan
NTG	New Technology Group
OFDM	orthogonal frequency-division multiple
OFDMA	orthogonal frequency-division multiple-access

PDA	personal digital assistants
PPDR	public protection & disaster relief
PSD	power spectral density
PU	authorized/licensed/primary user
PUE	primary user emulation
RBS	received BS
SACFA	Standing Advisory Committee on Radio Frequency Allocation
SDR	software-defined radio
SHV TV	super Hi-vision television
SU	unlicensed/secondary user
TRAI	Telecom Regulatory Authority of India
TVWS	television white space
UHDTV	ultra-HDTV
WPC	wireless planning and coordination
WRAN	wireless regional network
WS	white space

1

Novel Application of TV White Space

1.1 Introduction

Around the globe, there is a coordinated move for DSO by discontinuing analogue TV transmission. TV switchover to full digital broadcast service created
new spectrum opportunities due to higher spectrum efficiency compared to
analogue services. This DD consolidates "White Spaces" (WS)-"leftover"
frequencies resulting from the change of TV broadcasting from analogue to
digital [1].

There are a number of TV frequency bands that have been allocated for
terrestrial television broadcasting, but they have not been assigned to the
provision of television services in a particular licensed area. These unused
channels have traditionally served variety of purposes, arising from (i) the
need for guard bands between analogue TV services in the same license
area, (ii) the need for geographic separation between TV services that are in
different licensed areas but are broadcasting on the same channel (geograph-
ically interleaved spectrum bands) and (iii) usage opportunities in the areas
where channels are not allocated to broadcasters due to supply constraints
(small number of authorized or deployed services) or demand weakness
(e.g. low population density like rural areas or alternative transmission
technologies) [2].

The cleared bands due to DD and unused TV channels are called as
TVWSs, which provide an opportunity for CR technology and for deploying
new wireless services. In Chapter 2, it will be discussed in detail about CR
technology. Survey of spectrum utilization for 600–800 MHz band in India [3]
is similar to other countries. Plot for spectrum usage at two different locations
in India at a particular time for frequency band 600–800 MHz [3]. These plots
of spectrum utilization for 600–800 MHz show that spectrum in these bands
in India is underutilized and can be opened for CR applications and other
advanced new wireless services for next-generation networks.

In this section, the TVWS opportunities and regulation in India are discussed. Section 1.2 deals with the DD activities at international level including band plan for 698–806 MHz band for IMT applications. Section 1.3 gives a brief idea about regulatory framework in India. Section 1.4 elaborates the DD in India, and spectrum allocation in India for TV channels. Section 1.5 deals with joint task group (JTG), which has been constituted to study the compatibility between services and applications in 700 MHz band in India and also to develop a national channelling Plan in 700-MHz band. Section 1.6 discusses the opportunistic spectrum access in India focusing on the candidate frequencies for CR technology. Section 1.7 focuses on the various opportunities for exploiting TVWS in an efficient manner including IEEE 802.11af standards which is based on CR technology utilizing TVWS, proposed broadcasting services, femto cell for wireless broadband, public safety application. Section 1.8 discusses the applications in TVWS for rural India which helps in boosting the rural economy by providing e-governance, e-agriculture, e-learning, e-healthcare and e-animal husbandry. Section 1.9 elaborates the key use cases and scenarios for TVWS opportunities from rural India point of view. Finally, Section 1.10 gives a detail idea about the regulatory activities related to CR and TVWS.

1.2 DD: International Scenario

In the following, the DD at International level is studied. In USA, Federal Communications Commission (FCC) completed the auction of spectrum in 698–806 MHz called as 700 MHz band in March 2008 and final switchover to digital TV occurred on 12 June 2009 [4]. Following band plan is being considered for 698–806 MHz band for IMT applications.

- Arrangement segmented into two major bands with services related to mobile, high-power broadcasting and public safety.
- The maximum use of this arrangement with specifications in 3GPP is this 2 × 18 MHz + 2 × 20 MHz. 698–716 MHz U/L frequency paired with 728–746 MHz D/L and 758–768 MHz D/L paired with 788–798 MHz U/L.
- Actual 3GPP implementation is limited to band 13 (primarily Verizon) and band 17 (primarily AT&T).
- Inclusion of broadcasting and public safety and division into two segments [5].

In France, ARCEP (the French Electronic Communications and Postal Sector Regulator) commissioned a study in 2008 on the optimum framework for releasing available spectrum [6] on the development of the digital economy called Digital Plan 2012. The key findings are as follows:

- Allocating a proportion of the released spectrum for mobile broadband services adds greater value to the economy than if this band were allocated exclusively to digital TV services.
- Mobile broadband services will support political goals of "digital inclusion".
- It is vital that a detailed framework for the process of reallocating the DD spectrum is established as soon as possible.
- Comparison of the approach taken in other countries shows widespread international support for the sharing of DD spectrum.
- Licensing UMTS frequencies.
- Filling the gap in broadband coverage.
- Stimulating the deployment of Digital Terrestrial TV (DTTV) and reserving some of the DD generated by the switch-off of analogue broadcasting for the development of mobile broadband and very high-speed broadband.

In Europe, DD band is 790–862 MHz [7]. Following band plan is being considered at ECC for 790–862 MHz band for IMT applications.

- 790–820 MHz D/L frequency paired with 832–862 MHz U/L.
- 30 + 30 MHz FDD with 42 duplex separation and 820–832 MHz (12 MHz) as the centre gap.

Multitude of standards is currently under elaboration for operation in TVWS in various committees including ETSI [7], IEEE and ECMA. In UK, DSO is scheduled to complete in 2012. The Office of Communications (OFCOM) estimates that the digital switchover programme will release up to 112 MHz of spectrum band for new uses [8].

The digital switchover process is underway but a complete analogue switch-off is not easy. The full switch-off of analogue services can have terrible consequences if viewers were not adequately prepared. So, DD will require a careful planning and the involvement of the entire broadcast industry.

The process of analogue switch-off will differ in countries depending upon the market configuration. Table 1.1 indicates analogue switch-off (ASO) situation in various countries [8–21]. The countries that have not done full ASO can take useful lessons from those that have completed ASO about understanding best approaches, pitfalls that should be avoided, can help to ensure a successful process.

Table 1.1   ASO situation in various countries

Country	Completion of ASO
UK, Luxemburg, the Netherlands, Finland, Sweden, Germany, Belgium, Denmark, Estonia, France, Czech Republic, Croatia, Switzerland, Malta Slovenia, Japan, South Korea, Malta,	Completed
Bulgaria, Cyprus, Greece	At the end of 2012
Australia, New Zealand, South Africa	2013
Pan-Arab	2014
Poland, Slovakia, India, Russia, Tunisia, Albania, *Cambodia*	2015
Chile, Colombia	2017

In India, ASO on cable and terrestrial will have four phases, in a 3-year transition starting on 31 March 2012 and finishing on 31 March 2015 [22–24].

1.3 Regulatory Framework in India

In India, the Wireless Planning and Coordination (WPC) is the national radio regulatory authority responsible for frequency spectrum management. National Frequency Allocation Plan (NFAP) forms the basis for development and manufacturing of wireless equipment and spectrum utilization [25]. Currently, NFAP-2011 [26, 27] is finalized in which the vacant bands due to DD is allocated to other important services such as International Mobile Telecommunications (IMT) services data broadcasting, High definition TV (HDTV), Ultra HDTV (UHDTV), mobile TV services, Super Hi-Vision (SHV) TV and DTT.

1.4 DD: Indian Scenario

India has diverse population density, geographical area, tradition, affordability, livelihood, etc. It has specific spectrum needs to address the requirements. The only official broadcaster in India is "Doordarshan"; it covers almost the whole of the country. Satellite channels are mainly via the cable TV and DTH services. However, in rural India, the access to the satellite channels is little and hence Doordarshan has strong roots in this part of India. Doordarshan uses all the channels in the band 25–230 MHz for analogue operations. Normally, Doordarshan broadcast over two or three TV channels all over India. The second and the third channel are being broadcast for limited period. Other than Doordarshan, some rural telephony devices are also operating in this band, but now these are being replaced. Therefore, there is a strong possibility of availability of white spaces in TV band.

In India, the band between 698 and 806 MHz, most specifically called as 700 MHz band, is mainly used by TV broadcast services, at present. Digital broadcasting is roughly six times more efficient than analogue, allowing more channels to be carried across fewer airwaves. Plans that include the aspects of utilization of digital switchover allows for an increase in the efficiency, offering real opportunities for wireless innovation.

1.4.1 Spectrum Allocation in India

According to [27], the frequency band 470–806 MHz has been allotted to fixed, mobile broadcasting services on primary basis, and spectrum allocation for terrestrial broadcasting services. In India, 470–806 MHz band has been allotted to fixed, mobile broadcasting services on primary basis. In **UHF Band IV** (470–582 MHz), 14 TV channels, each with 8 MHz bandwidth, are available. Doordarshan is only the Government broadcaster that operates digital transmitters in four metros in this band. In **UHF Band V** (582–806 MHz), 28 TV channels each with 8 MHz bandwidth are available. Defence and BSNL are operating point-to-point microwave links in 610–806 MHz. Public Protection and Disaster Relief (PPDR) has some spots earmarked in 750–806 MHz. The UHF Band V above 806 MHz is also shared with other users of spectrum such as fixed and mobile services for transmission of data/voice and video (see Table 1.2). The complete switchover to digital transmission is a very challenging task in India considering huge analogue TV sets in rural India and in more populated parts of India but it will be completed in 2015.

Telecom Regulatory Authority of India (TRAI) recommended that for the growth of rural areas, Department of Telecommunication (DoT) should allocate this band for use by advanced wireless technologies for rural connectivity. DoT's guidelines for auction and allotment of spectrum for broadband wireless access (BWA) services and spectrum in 700 MHz bands shall be auctioned as and when it becomes available [26]. A complete switchover to digital transmission is a very challenging task in India considering huge analogue TV sets in rural India and in more populated parts of India. This will be slow process and hence Doordarshan suggested simulcast of analogue and digital transmission till complete switch-off of analogue transmission [28].

Table 1.2   Spectrum allocations for TV broadcasting/other services

Band	Spectrum	Number of TV Channels Available in Analogue Mode/Other Services	TV Channel Number
UHF Band IV	470–582 MHz	14	21–34
		Mobile TV using DVB-H	26
	582–806 MHz	28	35–62
	806–960 MHz	–	–
UHF Band V	610–806 MHz	BSNL, Defence operate point-to-point microwave links	
	746–806 MHz	Public Protection and Disaster Relief (PPDR)	
	Above 806 MHz	Fixed, mobile services for transmission of data/voice, cellular mobile services	

1.5 Joint Task Group and 700 MHz

In India, JTG has been constituted to study the compatibility between services/applications in 700 MHz band and also to develop a national channelling Plan in 700 MHz band. The Doordarshan [28], Ericsson [29], WIMAX Forum [30], TEMA [31], COAI [32] and Nokia Siemens [33] submitted a proposal to JTG about the utilization of 700 MHz band for IMT applications along with the band plan. NFAP-2011 includes the aspects of utilization of DD for IMT services, broadband services, mobile TV, DTT, data broadcasting and HDTV [26]. Active participation of Indian delegation at the WP5D meeting and previous AWF meetings leads to finalization of the following band plan by WP5D for region 3 countries [34]. India is not supporting mixed FDD and TDD mixed band plan, as such mix would lead to significant wastage of spectrum in guard bands to avoid FDD/TDD interference implications. Figure 1.1 indicates the band plan for 698–806 MHz for region 3 countries employing two options. Option 1 deals with FDD arrangement and option 2 deals with harmonized TDD arrangement for 698–806 MHz.

Option 1: FDD Arrangement for 698–806 MHz band:

- 45 + 45 MHz FDD arrangement with 10 MHz central gap (in brief, 2×45 MHz band plan). Conventional duplex arrangement, where the mobile transmit function is allocated to the lower spectrum block of

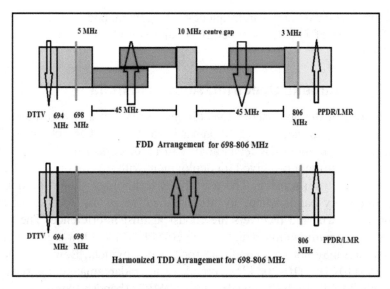

Figure 1.1 Band plan for region 3 countries [34].

the FDD pair (the earlier proposed reverse duplex arrangement, based on technical studies, had to be modified to conventional, to safeguard interference issues with GNSS).

- Dual duplexer arrangement.
- Internal guard band at 698 MHz = 5 MHz.
- Internal guard band at 806 MHz = 3 MHz.
- Additional 4 MHz external guard band (below 698 MHz deploying 8 MHz TV channel raster as in India). [A few countries deploying 6-MHz channel would need to provide the additional external guard band (4 MHz) at 698 MHz level internally in the band 698–806 MHz.]

Option 2: Harmonized TDD arrangement of 698–806 MHz band:

- The proposed TDD option frequency arrangements in the band 698–806 MHz has also been considered with appropriate guard bands on both sides as shown in Figure 1.1.
- This arrangement includes an internal guard band of 5 MHz at 698 MHz and another internal guard band of 3 MHz at 806 MHz.
- Additional "internal" guard bands may also be required between unsynchronized TDD networks.
- Additional 4 MHz external guard band below 698 MHz is also available because of the 8 MHz TV channel raster in India.

These two finalized band plan/frequency arrangement had since been sent by WP5D to SG5 of ITU (November 2011) and would now be considered by WRC-12.

1.6 Opportunistic Spectrum Access in India

CRs are one of the approaches for utilizing the WS in a shared manner. In India, presently no specific regulation has been made for CR technology. The detail discussion about CR technology will be discussed in Chapter 2. CR technology is still not perceived for deployment with several open challenges including spectrum sensing. The use of CR technology should be done in those frequency bands where spectrum utilization is low, location of base station is known, and receivers are robust against interference. The bands where spectrum utilization is high like in GSM/CDMA or low signal strength like satellite may not be a candidate for CR deployment. Lower frequency bands (9–110 kHz, 110–200 kHz), broadcasting, radar, amateur, radio paging and Public Mobile Radio Trunked systems (PMRTs) bands are most prominent candidate for CR technology as the activities in these bands are low and these services are almost outdated.

As shown in [27], the frequency bands identified by WRC-07 for use by IMT application on regional basis can also be explored for CR. They are 790–862 MHz (Regions 1 and 3), 698–790 MHz (in Regions 2 and 9 countries; and in Region 3 the following countries: Bangladesh, China, Republic of Korea, India, Japan, New Zealand, Papua New Guinea, the Philippines and Singapore), 3400–3600 MHz (over 80 administrations in Region 1 plus 9 countries in Region 3 including India, China, Japan and Republic of Korea). To ensure the adequate protection of services to which the frequency band is allocated, WRC-07 developed WRC-11 agenda item 1.17 to consider the results of sharing studies between the mobile services and other services in the band 790–862 MHz in Regions 1 and 3, in accordance with resolution 749 (WRC-07). Spectrum usage efficiency in IMT bands by primary services is not better, and most of the times some part of the spectrum are lying vacant or unused. The key benefit for deploying CR system in IMT bands is to improve the overall spectrum efficiency and increased flexibility.

According to NFAP-2011 [27], some of the bands that have been exempted from licensing requirements are also candidates for CR deployment:

- Utilization of very low power devices, in frequency band 50–200 kHz on non-interference, non-protection and shared (non-exclusive) basis (IND01 of NFAP 2011 [27]).

- Utilization of wireless equipments intended to be used while in motion or during halts, in the frequency band 26.957–27.283 MHz, with a maximum effective radiated power (ERP) of 5 Watts (IND10 of NFAP 2011 [27]).
- Utilization of low-power equipments for the remote control of cranes using frequencies 335.775, 335.7375, 335.7625, 335.7875, 335.875 and 335.8375 MHz, with a channel bandwidth of 10 kHz and maximum transmit power of 1 mW (IND19 of NFAP 2011 [27]).
- Utilization of very low-power remote cardiac monitoring RF wireless medical devices, medical implant communication/telemetry systems and other such medical RF wireless devices in frequency band 402–405 MHz using a maximum radiated power of 25 microWatts or less with channel emission bandwidth in 300 kHz (IND31 of NFAP 2011 [27]).
- Utilization of low-power RFID equipments or any other low-power wireless devices or equipments in the frequency band 865–867 MHz with a maximum transmitter power of 1 Watt (4 Watts ERP) with 200 kHz carrier band width (IND44 of NFAP 2011 [27]).
- Utilization of low-power equipments in the frequency band 2.4–2.4835 GHz using a maximum transmitter output power of 1 Watt (4 Watts ERP) with spectrum spread of 10 MHz or higher (IND62 of NFAP 2011 [27]).
- Utilization of low-power equipments in the frequency band 5.825–5.875 GHz using a maximum transmitter output power of 1 Watt (4 Watts ERP) with spectrum spread of 10 MHz or higher (IND72 of NFAP 2011 [27]).
- Utilization of low-power equipments for wireless access systems including radio local area networks, in the frequency band 5.21–5.350 GHz and 5.725–5.875 GHz using a maximum mean effective isotropic radiated power (EIRP) of 200 mW and a maximum mean effective isotropic radiated power density of 10 mW/MHz in any 1 MHz bandwidth, for the indoor applications (IND 67 of NFAP 2011 [27]).

1.7 Opportunities in TVWS

TVWSs provide an opportunity for CR technology and for deploying new wireless services. Spectrum regulatory bodies in various countries are studying the pros and cons of CR devices. Some countries have already made provisions for CR. FCC has already made provision for the use of CR device in TV bands

[35]. OFCOM from UK has also studied CR in their spectrum framework review and made provision for the use of unlicensed cognitive devices in TV band by using TVWS [36]. The Confederation of European Post and Telecommunication Agencies (CEPT) has implemented CR-based device to operate in TV band exploiting TVWS [37]. In Australia, Australian Communications and Media Authority (ACMA) studied the DD of 698–820 MHz which are also of interest to white space proponents. The regulator set out its preferred approach to configuring the DD spectrum in a discussion paper that it released in October 2010. That approach would result in a 9 MHz guard band from 694 to 703 MHz to prevent interference between high-power broadcasting services below 694 MHz and the mobile telecommunications networks most likely to be deployed above 703 MHz. This 9 MHz guard band will clearly be of interest to CR technology utilizing TVWS [38].

There are various opportunities for exploiting TVWS in an efficient manner. Some of them are as follows:

1.7.1 Wide Area Coverage in Rural Areas (e.g. IEEE 802.22)

IEEE 802.22 WRAN aims to provide a fixed wireless access with a typical cell radius of 33 km with effective isotropic radiated power (EIRP) of 4 W and a maximum of 100 km in rural and remote areas using CR technology in TVWS [39]. The spectral efficiency of 802.22 systems ranges from 0.5 to 5 bit/s/Hz. At an average value of 3 bit/s/Hz, the capacity of a 6 MHz channel can reach 18 Mbit/s. Broadband to rural areas might therefore be provided at a low cost. The use of IEEE 802.22-standard-based WRAN has already been allowed in most of the countries. Equipment in the market is also available.

1.7.2 Super Wi-Fi/Low-Power Broadband (e.g. IEEE 802.11af)

Wi-Fi has been successful and cost-effective technology for wireless internet access delivery across places. There are few spectrum bands that are reserved for industrial, scientific and medical purposes. These bands are de-licensed and referred to collectively as the industrial, scientific and medical (ISM) radio band [40]. This ISM band is mostly used for data communication for wireless internet access across places. There have been several extensions to increase the range and capacity of Wi-Fi. Since the much-used 2.4 GHz band for Wi-Fi is quite congested, there is a major concern with the interferences from home appliances and devices working in ISM band. The concept behind IEEE 802.11af [41] is to implement the wireless broadband networks by utilizing the bandwidth within TV broadcast stations, called super-Wi-Fi or "White-Fi" because of its extended capability over standard Wi-Fi use cases, "Super" – because of its cognitive properties, and "White" – due to work in a range of free TVWS frequencies.

In some urban areas like Asia, Europe and USA, there is no much white space (WS) for high-power systems such as fixed unlicensed devices that would operate from a fixed location and could be used to provide commercial services such as wireless broadband Internet access, but there could be a potential for low-power broadband systems like "portable/personal" unlicensed devices, such as wireless in-home local area networks (LANs) or Wi-Fi–like cards in laptop computers that exploit smaller portions of TVWS.

The development of IEEE 802.11af standard is more systematic with extended coverage and managed interference but it is not finalized. Promising applications with respect to Indian scenario are as follows:

- Traffic monitoring and precautionary measure for prevention from accident: Road accidents have earned India a dubious distinction. In India, over 1,000 deaths are recorded annually. In this respect, India has overtaken China and now reportedly it has the worst road traffic accident rate worldwide [42]. Traffic cameras can be installed at all major accident-prone intersections to provide real-time traffic monitoring. It will help the department of transportation to reduce congestion, travel time, fuel consumption, and also support local law enforcement as majority of the accidents take place on expressway.
- Anti-theft measure: Theft such as car theft, motorcycle theft, bicycle theft and personal theft are serious issues in India. Anti-theft measure is a wireless camera to police for surveillance. Also radios can be installed in city parks so that free Wi-Fi access can be provided to residents and city workers for surveillance.
- Remote monitoring of heavy wetland areas: The city and county can use the white space network to remotely monitor and manage heavy wetland areas such as north-east state of India (Assam, Mizoram, Nagaland, Manipur and Sikkim) to comply with Environmental Protection Agency (EPA, protecting human health and environment) regulations because these areas are hard to get.
- Remote monitoring of power plant: Power plants are usually far away like hydro power plants (Pophali) near the dams or nuclear power plants (such as Tarapore). Smart Grid technologies is a digitally enabled electrical grid that gathers, distributes and acts on information about the

behaviour of suppliers and consumers in order to improve the efficiency, reliability, economics and sustainability of electricity services. Smart Grid technology that utilizes TVWS could devise more efficient ways in order to improve system control, manage its supply and demand of electrical power and provide broadband Internet access to under-served areas (such as tribal villages in Himachal Pradesh).

- Providing temporary connectivity to manage large people assembled for exhibitions, festivals, election campaign.
- Community awareness sessions, medical campaigns.
- Backhaul for Wi-Fi in campuses, businesses, hotels, theatres (Wi-Fi could be IEEE 802.11af or IEEE 802.11a/b/g/n). The costly, time-consuming and challenging cabling issues can be avoided by utilizing TVWS as backhaul for Wi-Fi access points.
- The network could also be used to provide applications, such as expanded Internet connectivity for local schools, medical monitoring and other environmental monitoring.
- Rural broadband access: It is possible to have e-health, e-governance, e-learning, e-agriculture and e-animal husbandry applications to uplift the life and rural economy of India by providing broadband wireless access utilizing TVWS.
- Healthcare applications, vehicular communications and cellular networks offloading, and ad hoc networks (e.g. internal sensor network or mesh network communication).

1.7.3 Broadcasting Services

Doordarshan has already submitted proposal [28] to JTG-INDIA on 698–806 MHz band about the allocation of many important broadcasting services as listed below apart from digital terrestrial television broadcasting and simulcast in this band:

- Enhancement of broadcasting services, e.g. Mobile TV services.
- Improved quality broadcasting services, e.g. HDTV services.
- Additional improved quality broadcasting services, e.g. UHDTV or SHVTV in future.
- Broadcast Ancillary Services (BAS).
- Broadcast Programme Making Services (BPS).

Doordarshan proposed allocation of this band for broadcasting applications only. The proposed channel plan for 8 MHz bandwidth required for terrestrial TV broadcasting in UHF band is explained in Table 1.3:

Table 1.3 Proposed channel plan by Doordarshan

Channel Number	Frequency (MHz)	Channel Number	Frequency (MHz)
Ch#49	694–702	Ch#56	750–758
Ch#50	702–710	Ch#57	758–766
Ch#51	710–718	Ch#58	766–774
Ch#52	718–726	Ch#59	774–782
Ch#53	726–734	Ch#60	782–790
Ch#54	734–742	Ch#61	790–798
Ch#55	742–750	Ch#62	798–806

1.7.4 DVB-H with Cognitive Access to TVWS

DVB-H could be utilized as an unlicensed secondary network together with licensed network such as DVB-T, operating in UHF bands. Mobile TV providers prefer this band as it provides balance in the antenna size and coverage. It can be integrated with car terminals, portable digital TV sets and handheld portable convergence terminals. Still, the big challenge with DVB-H by mobile operators is the issue of suitable business and revenue models. When implementing DVB-H based on CR, the following design considerations must be taken into account:

- Time slicing: In order to reduce the average power consumption of the terminal and to make possible smooth handovers. The time-slicing technique enables a considerable battery power-saving. Further, time-slicing allows soft handover if the receiver moves from network-to-network cell.
- An enhanced error protection scheme on the link layer to increase reception robustness for indoor and mobile contexts.
- In-depth symbol interleaver: For further improvement of the transmitted signal robustness in impulse noise conditions and mobile environment.

1.7.5 LTE Extension

GSM/3G/High-Speed Packet Access (HSPA) system with its set roadmap and proven capabilities forms an appealing ecosystem. It is the best solution for deployment as a next-generation cellular communication infrastructure and is well positioned to succeed in mobile broadband [43]. An LTE network at 700 MHz would provide following advantages:

- Better propagation characteristics: Longer range of propagation (compared to GSM 310 MHz, 2.1 GHz or 2.5 GHz bands for 3 G/BWA) ensures network deployments to cover larger areas with fewer base stations.
- Ideal for providing wireless service in low population density regions, such as rural India as majority of future growth expected from rural India.
- Less infrastructure-reduced costs.
- Reduced capital expenditure, which makes deployment in rural or high-cost regions economically viable.
  - An LTE network at 700 MHz would be 70% cheaper to deploy than an LTE network at 2.1 GHz-GSMA [43].
  - Two or three times as many less sites required for initial coverage at 700 MHz compared to 2.1 GHz or 2.5 GHz [43] (see Figure 1.2) (Table 1.4).

Figure 1.3 clearly indicates that selecting 700 MHz band for various applications will save the relative capital expenditure by two or three times as many less sites required for initial coverage at 700 MHz compared to 2.1 GHz or 2.5 GHz. LTE has considerable flexibility, scalable channel bandwidths from 1.4 MHz to 20 MHz using OFDMA with both TDD and FDD operation. This optimizes the use of radio spectrum by making use of new spectrum and re-farmed spectrum opportunities. Operators evolving to

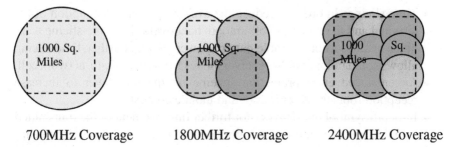

700MHz Coverage   1800MHz Coverage   2400MHz Coverage

Figure 1.2   Cell site coverage per thousand square miles [43] (1000 Sq.miles = 119.34 km).

Table 1.4   Comparative study of cost vs propagation at 700 MHz, 1900 MHz and 2.4 GHZ [43]

Cost	Propagation at Frequency		
	700 MHz	1900 MHz	2400 MHz
Total network cost @ 21 K/cell	$ 21,000	$ 600,000	$ 1,500,000
Network cost per customer	$ 31	$ 725	$ 1820
# Mos. to network cost breakeven	9 months	36 months	91 months

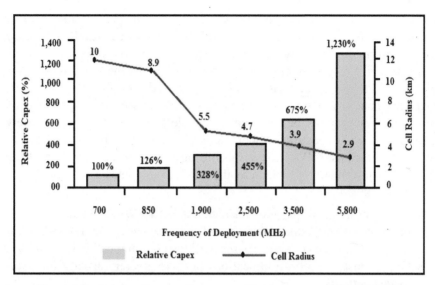

Figure 1.3 Relative Capex vs Frequency of Deployment vs Cell Radius [43].

LTE from GSM/WCDMA/HSPA will maintain full backward compatibility with legacy networks. The TVWS could also be exploited for managing peak traffic by obtaining and sharing channels on a temporary basis.

1.7.6 Femto Cell for Wireless Broadband in TVWS

The cellular mobile operators are considering femto cells deployed for indoor environment as a complement to microcells for the purpose of enhancing the coverage and capacity, as TVWS can penetrate through buildings and walls. In [44], the new concepts intra-operator spectrum WS reuse, multi-operator spectrum sharing and multi-service spectrum reuse exploit the spectrum of multiple operators and of multiple non-cellular services such as DTV broadcasts in femto cells deployed in a relatively isolated indoor environment. In this line, provisioning TVWS for femto cells can be an approach worth exploring.

1.7.7 Public Safety Application

Currently, all public safety communication infrastructures in India use narrow-band radios. The narrowband nature of these radios limits them to two-way voice communications with no inherent support for high-bandwidth transmission requirements such as interactive video communication, remote video surveillance of security or disaster sites, and do not provide the level of secure communication required by India's security forces resulting in easy leak of information to rogue entities, for example terrorists. One of the ways of providing safety systems is by using overlays to the existing systems such as TETRA-enhanced data services, optimized for efficient use of PMR frequency bands. Although it can be commercially deployed through WIMAX and LTE, they may not be sufficiently reliable as required by public safety measures. The CR spectrum broker can easily integrate a prioritizing mechanism that assigns TVWS for a specific disaster area based on spectrum availability information provided by the geo-location database.

CR use benefits are:

- Excellent signal propagation characteristic of TVWS where radio signals better penetrates building walls and floors when the individual radios move outside the coverage footprint of the communication system.
- Opportunistic spectrum allocations provide greater capacity for overloaded network.
- Dynamic reconfiguration and priority to better manage load and connectivity.

1.8 Empowering Rural India

About 80% of the Indian populations live in villages, and the agriculture is the main source of income. Indian agriculture is the largest contributor to the Indian economy, and it also plays an important role in the growth of socio-economic sectors of India. To uplift the people above poverty level, e-governance, e-agriculture, e-learning, e-healthcare and e-animal husbandry utilizing TVWS opportunities would help in decreasing farmer suicides, in decreasing primary school dropout rate, and in decreasing mortality rate by providing knowledge-based economy. Animal husbandry is a practical solution to bring farmers out of the poverty trap and its importance needs to be recognized in ICT programmes.

In India, there is stained stupendous growth for mobile. The rural subscriber base served by private GSM operators is approximately 75 million and growing about around 4–5 million every month, that is nearly 50% of the GSM subscriber added are from the rural areas [43]. Harmonization of 700 MHz will ensure that India use the same frequency to deploy the LTE, the next-generation mobile broadband technology. It will also provide significant social benefits, particularly in rural areas not served by fixed broadband.

The rural tele-density in the country is 18.97% as compared to urban tele-density as on 30 September 2009. In rural areas, BSNL has deployed Wireless in Local Loop (WLL) network to meet the demand of scattered and far-flung rural areas where connection of landline telephone is not techno-commercially feasible.

Key socio-economic goals identified by India's 11th Five-Year Plan 2007–2012 wrt MDG are mentioned below.

1.8.1 E-Agriculture

Agriculture is the only way to make a living for almost two-third of Indian population. Indian agriculture is the largest contributor to the Indian economy, and it is an integral part in the growth of socio-economic sectors of India. The e-agriculture helps in the following activities:

- Boosting agricultural GDP: timely information on weather forecasts and calamities.
- Better and spontaneous agricultural practices.
- Better marketing exposure and pricing.
- Reduction of agricultural risks and enhanced incomes.
- Better awareness and information: by subsidy in agriculture loans.
- Improved networking and communication.
- Facilitate online trading and e-commerce, better representation at various forums, authorities and platform, etc.
- Acquiring more area for cultivation purposes.
- Expanding irrigation facilities.
- Utilization of improved and advanced high-yielding variety of seeds.
- Implementing better techniques that uprisen from agriculture research.
- Water management.
- Plant protection activities through prudent utilization of pesticides, fertilizers and cropping applications.

1.8.2 E-Animal Husbandry

Animal husbandry is a practical solution to bring farmers out of the poverty trap and its importance needs to be recognized in ICT programmes. The livestock sector has vast potential in brining qualitative improvement in rural economy. Contribution of livestock sector in the total GDP during 2007–2008 was 4.4%. While reviewing the position of India as a largest producer of milk, Buddhikot et al. stated that milk is a key livestock commodity in India representing 67%

of total livestock output and 18% of total output by the agricultural sector [45]. Department of Animal Husbandry, Dairying & Fisheries (DAH&D) announces the schemes proposed to be implemented during 2010–2011 to improve GDP; among the proposed various schemes, one of the schemes for disease diagnosis and control is to introduce an online disease reporting system in the country in the form of National Animal Disease Reporting (NADR) system [45].

It is possible to implement NADR system using CR networks utilizing TVWS. In this proposed system, a dedicated computer network utilizing TVWS will be established linking each Taluka of the district to the district headquarter, each district of the state to the state headquarter and each state to the country's central unit. The information available at the Taluka would be electronically transmitted to the district veterinary office as well as the state office. Finally, from all the states the information would be compiled and transmitted to the centre at DAH&D. The computer linkages would reduce data transmission time as well as data compilation and report generation. With this system, farmers will be benefited by providing wireless connectivity among various units such as Taluka unit, district unit, state unit, centre unit, veterinary (Vet.) dispensary, Taluka level NGO. The central unit at DAH&D would be responsible for analysis and maintenance of data. All the disease-related information would be available online ensuring transparency and benefit of all concerned (Figure 1.4).

1.8.3 E-Health

It is possible to utilize TVWS for e-health application, viz. video conferencing. Through video conferencing doctors at a far-flung hospital will be able to diagnose the villagers for basic ailments and prescribe medicines online. The printout of the prescription will be available at the community service centres set up by the department and able to use the broadband connectivity to initiate an e-medicine programme wherein we will set up e-cardio testing and e-diagnosis facilities, thus providing following benefits:

- removing the lack of doctors in rural areas;
- removing language barriers through multi-lingual interfaces;
- bridging the lack of visual examination through video capability;
- reducing infant and maternal mortality;
- reducing malnutrition and anaemia among children and women; and

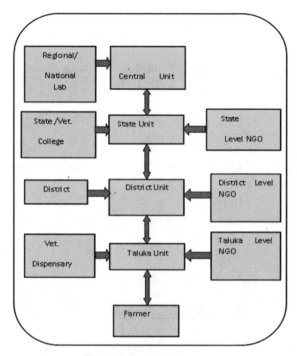

Figure 1.4   NADR system.

- boosting rural healthcare facilities: health awareness especially in case of snake bite, old age people health tracking—check-up timing, medicine timing, etc.

1.8.4 E-Education

In rural India, the parents seldom educate their children, and instead make their children work in the fields. Poverty, and lack of sufficient infrastructure and teaching staff, can be attributed to the lack of education in rural India. Also there is a genderwise discrimination for the education. According to [46], dropout rate in children's in the age of 5–14 years is shown in Tables 1.5 and 1.6.

By providing e-education facility in TVWS, it is possible to boost the education in rural area by:

- providing high bandwidth access for education courses;
- providing video-based and computer-based trainings (CBTs);
- reducing elementary school dropout rates;

Table 1.5   The drop-out rates for children aged (5–14 years)

(Percentage)

	Rural			Urban		
	Total	Male	Female	Total	Male	Female
Children not interested in studies	37.2	33.1	31.3	37.4	38.0	36.6
Parents not interested in studies	12.5	7.8	17.4	8.8	6.9	11.0
Unable to cope	16.4	1.7	8.1	13.7	13.0	14.5
To work for wage /salary	2.5	1.0	1.0	4.6	5.6	3.3
Participation in other economic activities	6.1	7.2	5.0	5.3	7.6	2.7
Attend to domestic duties	3.7	0.8	6.7	3.9	1.8	6.3
Financial constraints	11.2	12.0	10.4	15.8	15.7	16.0
Other reasons	7.9	0.9	9.8	7.4	8.7	6.0

Table 1.6   Percentage of out-of-school children (currently not attending school) 5–24 years (in percentage)

Out-of-School Children (5–14 years)	Percentage Dropout (Enrolled But Currently Not Attending School)			Never Enrolled		
	Both	Male	Female	Both	Male	Female
Rural	17.6	19.2	16.5	82.3	80.3	83.5
Urban	39.3	42.3	36.6	60.1	57.7	63.4

- boosting enrolment for higher education: rate in higher education in rural areas was 8.40 percent; and
- increasing adult, specifically female, literacy.

1.8.5 E-Governance

- Remove bureaucracy by providing instant reference of land and other relevant records, etc.
- Provide government information such as policies, forms, schemes.
- Check agricultural market prices online.
- Issue the various certificates such as Birth and Death certificates.

1.9 Use Cases for TVWS Usage

The ETSI RRS technical committee [7] is currently dealing with TVWS-related standardization focusing on both cellular and short-range type of applications. In sequel, the key use cases and scenarios are presented from rural India point of view [2]. These key use cases can be implemented depending on users and BS geo-location and user's mobility.

1.9.1 Use Case: Mid-/Long-Range Wireless Access

Internet access is provided by BS to the end-user by exploiting TVWS over the ranges similar to the cellular system 0–10 km. depending on user's mobility; it is further classified as follows.

1.9.1.1 Mid-/long-range: no mobility

Wireless access is provided from BS towards fixed mounted home access point/base station as shown in Figure 1.5. The geo-location for both BS and fixed device are well known.

1.9.1.2 Mid-/long-range: low mobility

In this scenario, wireless access is provided from BS towards mobile devices where the user has low mobility, for example taking round in farm, walking. Sensing results for primary/licensed users (PU) fetched for the current location

Communication in TVWS Frequency bands

Wi-Fi

Figure 1.5 Mid-/long-range wireless access: no mobility (Konkan house).

are not getting invalid because of its low mobility. The geo-location from mobile devices must be determined during operation. Cellular positioning system or GPS can be utilized to keep a track of mobile user use. The geo-location of mobile user from BS is well known.

1.9.1.3 Mid-/long range: high mobility

In this case, wireless access is provided from a BS towards mobile devices and where the mobile user is moving fast, for example, in the vehicle such as car or train. The geo-location from BS is well known. In this case, the sensing results for primary users fetched from the current location are getting invalid quickly because of fast mobility of the user. More efforts need to be taken to exploiting TVWS as this use case as it sets high constraints on PU sensing. It can be questioned if high mobility will be supported in TVWSs at all.

1.9.1.4 Centralized network management

A network centric solution in regard to allocating available TVWS for the user terminal to get connectivity can be considered. The available TVWS is considered based on location rather than in time. The TVWS would be largely available in rural area of India and in time. Once the terminal accesses the network it can be left under the control of the network, high-layer signalling can be utilized for this purpose, for example system broadcast messages or handover command to handoff to a new frequency can be used to notify terminals about the change in the frequency.

1.9.2 Use Case: Short-Range Wireless Access

Internet access is provided by BS to the end user by utilizing TVWS over the ranges similar to the cellular system 0–50 m. This use case is further studied as follows:

1.9.2.1 Uncoordinated networks

Multiple coordinated networks can access WS frequency bands. This is happening in rural area where the houses are big, each residents equipped with their own access points operating (AP) in WS frequency band. This type of use case is useful for surfing at the time of farming, diagnosis, in Animal husbandry, etc. (Figure 1.6).

1.9.2.2 Coordinated networks

WS networks in the close proximity are operated in coordinated manner by WS operator. Examples of this kind of usage can be academic institute, Thasildar office, Gram Panchayat office as shown in Figure 1.7. Gram

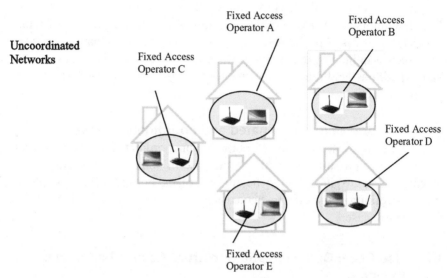

Figure 1.6 Uncoordinated networks, short range (Villages in Spiti, Himachal Pradesh, India).

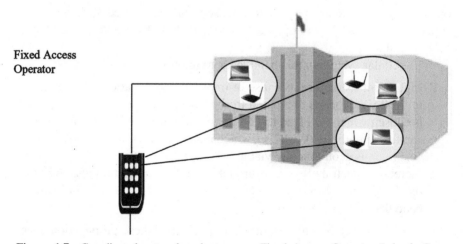

Figure 1.7 Coordinated networks, short range, Fixed Access Operator (school, Gram panchayat or Thasildar office).

Panchayat office, a local self-government at the village, is responsible for overall village administration such as looking after public health, education, keeping records of births–deaths, land details. Tahasildar office is a head of Taluka administration responsible for regulating the functionality of Gram Panchayat office is responsible for emergency services such as natural calamity case, such as lightening death, flood affect, revenue collection such as rehabilitation/petitions/land reforms/enactment of various legislations and all other technical functions.

1.9.2.3 Hybrid of uncoordinated and coordinated networks

Overall deployment consists of both the uncoordinated and coordinated networks. Such a situation could be the case in veterinary hospital, government general hospital (primary health centre) and house for older people. In order to work properly, effective coexistence methods need to be in place for this scenario case.

1.9.3 Use Case: Opportunistic Spectrum Access by Cellular Systems

In this case, TVWS slots are available sporadically for unlicensed/secondary users (SU) such as multi-mode user terminals being able to operate as, among other systems, cellular systems in licensed and unlicensed spectrum. The supported unlicensed spectrum is assumed to include TVWS, that is, 470–806 MHz in India.

The cellular system in this band leads to the following advantages:

1. Lighter infrastructure: because of improved propagation characteristics in the TVWS bands compared to typical license band, a large cell size is chosen which will lead to lighter infrastructure and results in reduced capital expenditure (CAPEX), which makes deployment in rural or high-cost regions economically viable.
2. Increased spectral efficiency through reduced propagation loss: A possible deployment choice is to keep a cell size as it is the case of licensed band deployment leads to:
   - Higher QoS achieved in a given cell. But these propagation characteristics may also increase interference issues which require an adequate handling (e.g. power management, suitable frequency re-use factor for TVWS.)

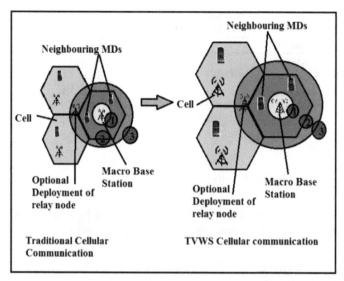

Figure 1.8 Improved coverage and reduced propagation loss in TVWS (note: symbols 1, 2, 3 indicate decreasing QoS, throughput level, etc.).

- Higher QoS within a given cell at a lower received BS (RBS)/mobile devices output power level. The inherent power consumption can be reduced.
- The hybrid solution of combination A and B is possible, that is a moderate reduction in RBS output power levels combined with a moderate improvement of the QoS.
- Increased spectral efficiency through extended macro diversity: A possible deployment choice is to keep cell size as it is the case of licensed band deployment. Then, joint operation of neighbouring RBS can be exploited in order to achieve a higher macro-diversity gain in UL or in DL (see Figure 1.8).

1.9.4 Use Case: Ad Hoc Networking over WS Frequency Bands

In this use case, the devices (user devices and other devices such as access points) can communicate with each other to share to execute other similar tasks. The communication happens by forming an ad hoc network operating on WS frequency band.

There can be two or more devices in the ad hoc network formed like device-to-device connectivity (video transfer in case of NADR system) and

Table 1.7   Scenarios and typical ranges in IEEE 802.11af standard based on CR networks utilizing TVWS

Scenarios	Ranges
Backhaul	10 km
LTE extension in TVWS	0–10 km
Rural broadband	0–10 km
Terminal to terminal Cellular	10–1000 m
Ad hoc network in TVWS	0–100 m
Femto cell in TVWS	0–100 m

ad hoc networking (e-health; the devices can be operating a localized social networking service). Thus, in Table 1.7 categorization of scenarios and typical ranges based on the above TVWS usage embedding different ranges, mobility requirements and QoS requirements in the CR networks is done.

1.10 QoS in TVWS

One of the challenging features of the TVWS is its variation across space and time. More specifically, the available channels are not contiguous and vary from one location to another. In addition, the white space available in a given location can vary with time if one or more of the TV band Pus start/stop operation. Opportunistic access to TVWS is interruptible in the sense that CR has to cease transmission immediately and relocate to a new band as soon as the DVB-T or PMSE which is already using white space on secondary basis appears. The delay associated with such relocations may face cognitive users with abrupt QoS degradation as communication peers need to coordinate the frequency transition, and many parameters across the protocol stack have to be reset to match the characteristics of the new frequency band. Therefore, cognitive radio links built on TVWS are inherently unreliable. The issue is on how to provide the best quality TVWS for secondary usage to maximize persistence of allocations while avoiding interference with primary users.

The QoS requirements for PMSE are as follows:

1.10.1 High QoS

The wireless microphones and remote controls (e.g. fireworks) need a high reliable radio link interface else would cause degradation in quality of the production. The maximum tolerable delay 1 min with required bandwidth of several of 100 kHz.

Table 1.8 Technical characteristics of PMSE

Applications	RF Output	Max. Bandwidth (kHz)	RF Transmission Height
PWMs	30 mw	200	1.5–12 m
Talkback	30 mw	20–50	1.5–12 m

1.10.2 Moderate QoS

The equipment can have a certain probability of interference and can tolerate certain interference than mobile speech services. These devices are used for speech that is not intended for broadcasting. The required bandwidth of 20–50 kHz is sufficient with tolerable delay not more than 1 min (Table 1.8).

1.11 Regulatory Activities Related to CR and TVWS

On the regulatory field, possible Indian approaches operating secondary TVWS need to study by national regulators WPC. More efforts should be given to the development of IEEE 802.11af. More efforts should be devoted to the assessment of 3 cognitive techniques such as sensing, geo-location database and beacon such that the selected sensing and geo-location techniques will provide protection to the licensed radio services. Geo-location must be present in all fixed devices, with an accuracy of +/–50 m. This position information is used to query a database for a list of available channels that can be used for cognitive devices operation. The database will include the information on all TV signals and may also have information on wireless microphone usage because low-power wireless microphones signal detection is challenging. The following aspects to be considered:

- Protection of digital broadcasting services including mobile TV.
- Protection of fixed and mobile services in the frequency band 470–520 MHz and 520–585 MHz.
- Protection of services adjacent to 470–806 MHz band, that is protection of mobile satellite services and broadcasting except aeronautical mobile satellite (R) service in the frequency band 806–890 MHz, and protection of IMT applications in the frequency band 450.5–457.5 MHz paired with 460.5–467.5 MHz.
- Protection PPDR communications.
- Definition of the requirements for the geo-location database approach.

As per [47–50] regulation requirements for IEEE 802.11af standard based on CR utilizing, TVWS is as shown in Table 1.9. The exact regulatory framework

Table 1.9   Regulation requirements for IEEE 802.11af

Parameter	[47]	[48]	[49]	[50]
TV signals (8 MHz) Quantized Signal energy	Not required	−114 dBm	Not required	Not required
Wireless microphone	Not required	−76 dBm in	Not required	Not required
Quantized Signal energy		200 kHz		
Geo-location accuracy	50 m	100 m	–	Not specified
Transmit power (Fixed) EIRP	1 W (with max 6 dBi antenna gain)	100 mW	–	Local Specific
Transmit power (portable) EIRP	100 mW	100 mW	–	Local Specific
Transmit Power in adjacent band to DTT signals	40 mW	20 mW	–	Local Specific
Out of band radiation	−55 dBm under the in-band level	–	–	–
Channel Availability Check	1 min	1 sec	–	–

of CR technology utilizing TVWS frequency band is yet to be finalized that guarantee an efficient exploitation of frequency bands in TVWS. As more research went along, it came clearer that geo-location will play the main role of defining the channels that must be free of wireless sensing device transmissions. The spectrum management and resource management of CR technology contain all the regulation constraints for all of the TVWS opportunities and that additional information will be taken from the network in order to exploit this portfolio to allocate the spectrum access using QoS-based rules.

There is a mandate given to by Indian government required from network operators a mandatory provision of connectivity to rural areas. Many operators find this difficult to address, because it is economically not attractive. Instead, they would prefer to pay the associated penalties. Using TVWS bands and the dynamic spectrum sharing, it is feasible to reduce the number of base stations providing the same or even higher throughputs that can encourage operators to extend their coverage in rural areas and ultimately uplift the rural area.

1.12 Conclusions

The DD of 700 MHz band offers real opportunities for wireless innovation. This 700 MHz band attracted many researchers and stakeholders for mobile communications by providing a rare opportunity to have cost-effective wireless solutions due to its excellent propagation characteristics compared to GSM 310 MHz, 2.1 GHz or 2.5 GHz bands for 3 G/BWA. The scope and nature of opportunities for WS created by DD (700 MHz band) in India has been discussed. The DD scenarios, regulatory aspects in India and the most promising applications of TVWS are discussed. NFAP-2011 leads the aspects of utilization of DD for IMT services, broadband services, mobile TV, Digital Terrestrial Television (DTT), data broadcasting and HDTV. The DD scenario is very beneficial to uplift the rural India by providing the e-education, e-health, e-agriculture and e-animal husbandry. Though there is only a single TV broadcast agency "Doordarshan" and they can easily take care of switching to digital transmission nationwide, there may be a backlash from rural population to switch. Most likely this will be done in several stages, according to the plans of Indian Government.

TVWS opportunities for rural area with the application such as e-education, e-agriculture, e-health and e-Animal husbandry are discussed, with special attention to e-animal husbandry application; NADR system based on CR technology for improving GDP is proposed. Also the use cases based on users and BS geo-location and user's mobility has been discussed. Regulation rules are set to prevent harmful interference to licensed users but do not provide insights to operate the available spectrum efficiently. Current research initiatives are tackling this objective.

References

[1] Dhope, T., D. Simunic, and R. Prasad. "TVWS opportunities and regulatory aspects in India," *14th International Symposium on WPMC '11*, Brest, France, pp. 566–570, Oct 2011.

[2] Dhope, T., D. Simunic, and R. Prasad. "TVWS opportunities and regulation: empowering rural India," *14th International Symposium on WPMC '11*, Brest, France, pp. 201–205, Oct 2011.

[3] Tripathi, P. S. M. Chandra, A. "Radio Spectrum Monitoring for CR," *2nd International conference on WIRELESSVITAE 2011*, pp. 1–5, Apr–Mar 2011.

[4] www.fcc.gov/700Mhzbandplan/2/7/2011

[5] www.publicsafetyfcc.gov/pshs/public-safety-specrum/700Mhz/release.htm section 2009/ 12/9/2009.

[6] http//ec.europa.eu/information_society/policy/ecomm/doc/implementation_enforcement/annualreports/14threport/fr.pdf

[7] ETSI reconfigurable Radio systems Technical Committee, information available at http://portal.etsi.org

[8] OFCOM, "DD: cognitive access," statement on licence exempting cognitive devices using interleaved spectrum, July 2009, Available: http://www.ofcom.org.uk/consult/condocs/cognitive/statement/statement.pdf

[9] EU Member States on course for analogue terrestrial TV switch off.

[10] Europa.eu/rapid/pressReleasesActionDo?reference=IP/09/266Brussels, 16th Feb 2009.

[11] http://www.dvb.org/about_dvb/dvb_worldwide/denmark/, 2nd November 2009.

[12] http://www.dvb.org/about_dvb/dvb_worldwide/estonia/, 31st August 2010.

[13] http://www.dvb.org/about_dvb/dvb_worldwide/spain/

[14] http://www.dvb.org/about_dvb/dvb_worldwide/slovenia/, 27th December 2011.

[15] http://www.dvb.org/about_dvb/dvb_worldwide/malta/, 17th October 2011.

[16] http://www.dvb.org/about_dvb/dvb_worldwide/austria/, 6th June 2011.

[17] http://www.dvb.org/about_dvb/dvb_worldwide/croatia/, 22nd November 2010 ACMA, "Realising the DD - Direction 2010," Minister for Broadband, Communications and the Digital Economy, 9 July 2010, Accessed 15 November 2010 at http://tinyurl.com/2dg69en, 2010.

[18] "When is my area going digital? – Going Digital," New Zealand Ministry for Culture and Heritage. http://www.goingdigital.co.nz/making-the-switch/coverage-areas-2/coverage-areas.html. Retrieved 17th July 2011.

[19] "Broadcasting Digitization Schedule," DPA: The Association for Promotion of Digital Broadcasting. http://www.dpa.or.jp/english/schedule/index.html. Retrieved 16th November 2009.

[20] http://www.dvb.org/about_dvb/dvb_worldwide/tunisia/

[21] www.content-technology.com/asiapacificnews/p=20 Content + Technology: DVB-T2 Trialled in Malaysia, Retrieved on 10th June 2011.

[22] www.MDA.gov.sg/public/DigitalTV/pages/IntroductionDigitalTV.aspx Media Development Authority Introduction to Digital TV http://www.dvb.org/about_dvb/dvb_worldwide/taiwan/, 1st Aug 2011.

[23] http://www.trai.gov.in/WriteReadData/trai/upload/Recommendations/79/laterdigital23feb11.pdf, 22nd Feb. 2011.

[24] Government of India, WPC wing, Ministry of Communications and Information Technology, DoT http://210.212.79.13/ http://www.wpc.dot.gov.in/Homepage

[25] National Frequency Allocation Table http://www.wpc.dot.gov.in/Doc Files/DraftNationalFrequencyAllocationTable-2011.pdf

[26] Draft National Frequency Allocation Plan-2011 http://www.wpc.dot.gov.in/DocFiles/DraftchannellingPlanforNFAT-2011.pdf

[27] TRAI consultation paper Consultation Paper on Issues Relating to Mobile Television Service 9/2007, http://www.trai.gov.in/Write Read-Data/trai/upload/ConsultationPapers/121/cpaper18thsep07.pdf

[28] Draft India Remarks in the National Frequency Allocation Table, http://www.wpc.dot.gov.in/DocFiles/DrafINDremarksforNFAP-2011.pdf

[29] www.wpc.dot.gov.in/DocFiles/ProposalfromDoordarshantoJTG-India.doc, 30th Sept. 2011.

[30] www.wpc.dot.gov.in/DocFiles/EricssonproposalforUHFbandinIndia3.pdf 17th Aug. 2009.

[31] www.wpc.dot.gov.in/DocFiles/ProposalfromWiMAXForumIndia.doc, 2009.

[32] www.wpc.dot.gov.in/DocFiles/ProposalfromTEMA.doc, 2009.

[33] www.wpc.dot.gov.in/DocFiles/ProposalfromCOAI.doc, 2009.

[34] www.wpc.dot.gov.in/DocFiles/Proposalfromnokia-siemens.doc August 2009.

[35] T. R. Dua, http://www.gisfi.org/wg_documentsGISFI_SPCT_2010743.pdf, Dec. 2010.

[36] FCC Spectrum Policy Task Force, ET Docket No. 04-186, Nov. 2008, www.fcc.org

[37] Ofcom: DigitalDevidend:clearing the 800 MHz band, http://www.ofcom.org.uk/consult/condocs/cognitive/

[38] Report C from CEPT to the European Commission, July 2008, www.erodocdb.dk/Doics/doc98/official/pdf/CEPTREP024.PDF

[39] ACMA, "Spectrum Reallocation in the 700 MHz DD Band," Australian Communications and Media Authority, Commonwealth of Australia,

Canberra, Accessed 2 December 2010 at, 'http://tinyurl.com/22uddg7', 2010.

[40] 802.22 Working Group, "IEEE 802.22 D1: draft standard for wireless regional area networks," March 2008, http://grouper.ieee.org/groups/802/22/

[41] Ekram Hossain, Dusit Niya, Zhu Han. Dynamic Spectrum Access and Management in CR Networks, Cambridge University Press, ISBN-13 978-0-511-58032-1, 2009.

[42] 802.11 Working Group, IEEE P802.11af D0.06: Draft Standard for Information Technology, Oct. 2010, http://www.ieee802.org/11/

[43] http://www.indiastat.com/crimeandlaw/6/accidents/35/roadaccidents/2997/stats.aspx

[44] Gupta, Sudhir. Use of 700 MHz-TRAI http://www.cmai.org.inspptUse of700MHz-TRAI.pdf4th Sept 2009.

[45] M. M. Buddhikot, I. Kennedy, F. Mullany, and H. Viswanathan, "Ultra-Broadband Femtocells via Opportunistic Reuse of Multi-Operator and Multi-Service Spectrum," Bell Labs Technical Journal, pp. 79–14, 2009.

[46] http://dahd.nic.in/annualplan/Coverpage.doc, 2010.

[47] Jayachandran, U. "How High Are Dropout Rates in India?" Economic and Political Weekly, pp. 982–983, Mar. 2007.

[48] FCC Second Memorandum Opinion and Order, September 2010, http://www.fcc.gov/

[49] DD: cognitive access Consultation on licence-exempting cognitive devices using interleaved spectrum, May 2009.

[50] William Webb "Cognitive access to the interleaved channels: Update and next steps," The IET seminar on CR Communication, London, Oct. 2010.

[51] CEPT WG SE43, Draft Report ECC 10, "Technical and operational requirements for the possible operation of CR systems in the 'white spaces' of the frequency band 470–790 MHz," available at http://www.ero.dk

2

Spectrum Sensing in Cognitive Radio

2.1 Introduction

In our day-to-day life, overwhelming amount of devices such as personal digital assistants (PDA), TV remote controls, cellular phones, satellite TV receivers, and mobile computers are based on wireless communications technology. The complexity of wireless system requires a careful design, especially related to bandwidth and energy efficiency. The energy efficiency is getting more and more on importance, due to increasing penetration of various wireless systems in different battery-oriented applications, as well as due to the more conscious global view on the need for "greening the Earth". Many more new technologies are emerging, which demand more spectrum and bandwidth for faster growth. Bandwidth efficiency is very important parameter, because it relates to frequency spectrum, which is naturally limited resource. It is not possible to get new spectrum allocation without the international coordination on the global level. Therefore, efficient use of existing spectrum is of prime interest as a research objective. CR has been proposed to meet the ever-increasing demand of the radio spectrum by allocating the spectrum dynamically without causing interference to the licensed users [1–11].

Detail explanation of dynamic spectrum access is given in Section 2.1. Section 2.2 elaborates what is CR, different definitions of CR described by various committees; how the CR is different from traditional radio and software-defined radio (SDR), what are the different capabilities of CR, and what is cognitive cycle In order to utilize the spectrum efficiently, CR needs to sense the spectrum frequently which makes spectrum sensing a crucial task and is an important part of Green Engineering. Section 2.3 overviews the various spectrum sensing challenges. The characteristics of spectrum sensing are described in Section 2.4. Section 2.5 gives detailed study of different types of spectrum sensing algorithm together with advantages and disadvantages of each method.

Energy detection method is a basic method for spectrum sensing, which requires knowledge of noise power. Covariance-based detection exploits space–time signal correlation that does not require the knowledge of noise and signal power. The performance of various spectrum sensing algorithms such as energy detection, covariance-based detection, and hybrid detection method is considered for IEEE 802.11af standard, which is based on CR utilizing TVWS.

Section 2.6 elaborates the analysis of energy detection and covariance absolute value method which is followed by simulation results in Section 2.6.3 for detection of PU. The PU is DVB-T TV signal in 2K mode with 8 MHz bandwidth considered. This 2K mode is intended for mobile reception of standard definition TV signal according to the European Telecommunications Standards Institute (ETSI) specifications under different time variant fading channels such as Rician and Rayleigh. The energy detection method relies on estimation of noise power for signal detection; therefore, the performance of energy detection with noise uncertainty is studied in Section 2.6.4 and compared with the covariance-based detection method. The need for hybrid detection algorithm is identified by comparing energy detection and covariance-based detection in Section 2.7. The simulation results for hybrid detection method for detection of PU under different time variant fading channels such as Rician and Rayleigh are illustrated in Section 2.7.4.

2.2 Dynamic Spectrum Access

The current spectrum allocation process operates at national, regional, and international level. At international level, International Telecommunication Union (ITU) is responsible for spectrum management [1]. Broadly, international bodies tend to set out high-level guidance which national bodies adhere to in setting more detailed policy. International coordination is essential in some cases because the zones of possible interference extend beyond national geographical boundaries as users are inherently international, for example, aviation. The current widely used method of allocation of spectrum is known as "command and control". In this method, radio spectrum is divided into spectrum bands that are allocated to specific technology-based services, such as mobile, fixed, broadcast, fixed satellite, and mobile satellite services on exclusive basis. This command-and-control-based spectrum management framework guarantees that the radio frequency spectrum will be exclusively licensed to a license (authorized/primary) user and can use the spectrum without any interference. This traditional approach of spectrum management justified successfully to avoid interference and played effective role in expansion of few services such as GSM worldwide through coordination and harmonization.

The existing static approach of spectrum management model is not efficient to cater the present and future requirement of spectrum for new wireless applications. The concept of dynamic spectrum access (DSA) has emerged dramatically for improving spectrum utilization. Some steps, such as delicensing more spectrum bands and allowing spectrum trading/sharing, toward dynamic spectrum were implemented by various administrations.

The spectrum utilization study in India reported that spectrum is underutilized in frequency bands (100–300 MHz) at given time and location as shown in Figures 1.1 and 1.2 of Chapter 1. These plots of spectrum utilization for 100–300 MHz show that spectrum in these bands in India is underutilized and can be opened for CR applications which is based on the concept of dynamic spectrum management for next-generation networks.

DSA can be defined [4] as a mechanism adapted to adjust the spectrum resource usage in a near-real-time manner in response to the changing environment and objective (e.g., type of application and available channel), changes of radio state (e.g., transmission mode, location, and battery status), and changes in environment and external constraints (e.g., radio propagation, operational policy). In simple way, DSA allows SU to access spectrum which has already been allocated to PU. The key characteristic of DSA is their ability to exploit knowledge of their electromagnetic environment to adapt their operation and access to spectrum. The opportunistic spectrum access is another name given to dynamic spectrum access and is an integral part of the larger concept of CRs. DSA is basically divided into four major phases [4]:

- *Spectrum sensing*: Spectrum sensing is to identify the spectrum opportunity that is unused part of the spectrum WS (i.e., it may be frequency band, time, location, and code) by periodically sensing the target frequency band and also determines the method of accessing it without interfering with the transmission PU.

- *Spectrum analysis*: Information from spectrum sensing is analyzed to gain knowledge about WS by considering the parameters such as interference estimation, duration of availability, and probability of collision with a licensed user, and a decision to access the spectrum (e.g., frequency, time, bandwidth, type of modulation, transmit power, location, and code) is made by optimizing the system performance given the desired objective and constraints.

- *Spectrum access*: The WSs are accessed by the SUs depending on the outcome of spectrum analysis phase.
- *Spectrum handoff*: It is a function related to the change of operating frequency band when a PU starts accessing a radio channel which is currently being used by a SU; then, SU immediately changes its operating frequency band. Spectrum handoff ensures that the data transmission by the SU can continue in the new spectrum band.

According to [3] DSA, where it can be deployed, can be categorized into four different classes:

- First class belongs to spectrum that is assigned but rarely utilized within a specific geographic region.
- The second class belongs to fixed signals, where the spectrum appears to be fully utilized, but because of the predictable nature of the signal, time-based opportunities are available for CR.
- The third category belongs to those spectral regions that are infrequently utilized, where the CR must be capable in detecting incumbent/PU transmissions when they occur, and also strong in ceasing transmissions immediately or moving the transmissions to other unoccupied channel (overlay approach).
- The last category is those regions where the spectrum is relatively well utilized, but still has some capacity available.

The Wireless World Research Forum CR and Management of Spectrum and Radio Resources [4] redefined the above 4 approaches for opportunistic spectrum access into two as follows:

- Spectrum underlay: In underlay spectrum access, both PU and SU can simultaneously transmit as long as interference from SU does not degrade the quality-of-service (QoS) of PU. When SU used licensed band with conditions that operation of unlicensed user would not interfere with the licensed user is known as the **vertical sharing.** Spectrum underlay can be used for CR systems using CDMA or UWB technology. In spectrum underlay, defining interference level and how it would be distributed between SUs are major issues in its implementation. UWB communication relies on the fact that if the bandwidth is increased, then reliable data transmission can occur even at power levels so low that PUs in the same spectral bands are not affected. However, spreading transmission power equally across a wide bandwidth could be largely suboptimal in case of strong in-band interference. Figure 2.1 illustrates the transmitter power spectrum density profiles in UWB underlay spectrum sharing approach.

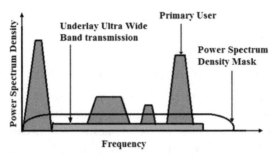

Figure 2.1 Transmit power allocation in UWB spectrum sharing.

- Spectrum overlay: In overlay spectrum access, SU is allowed to access the spectrum, which is allocated to some PUs when the spectrum is not used by PU. Access of WS is an example of spectrum overlay. Spectrum overlay can be used for CR in FDMA, TDMA, or OFDM wireless systems. This approach is compatible with the existing spectrum allocation methodology; therefore, the legacy systems can continue to operate without being affected by the CR users. Figure 2.2 illustrates the transmitter power spectrum density profiles in CR overlay spectrum sharing approach. However, identification of WS is a major hurdle in its implementation. Spectrum overlay is also known as the spectrum sharing. Spectrum sharing is not new a phenomenon. The unlicensed 2.4-GHz-frequency band in which WLAN, Wi-Fi, and Bluetooth technology operate is one of the best examples of spectrum sharing. Such spectrum sharing is also known as **horizontal sharing**. Unique to CR operation is the requirement to sense the environment over huge swaths of spectrum and adapt to it since the radio does not have primary rights to any

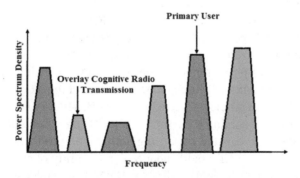

Figure 2.2 Transmitter power allocation in CR spectrum sharing.

preassigned frequencies. Note that CRs are allowed to transmit increased transmitter power levels but must not cause co-channel and adjacent channel interference to primary users in the vicinity.

DSA can be deployed in either centralized or distributed manner [4]:

- In centralized case, a central controller makes the decision on spectrum access by all SUs after collecting information about the spectrum usage of the PUS as well as information about the transmission requirements of the SUS. However, information collection and exchange to and from the central controller is a major issue in centralized system.
- In distributed dynamic spectrum access, SU can make a decision on spectrum access independently and autonomously. Since each SU has to collect information about the ambient radio environment and make its decision locally, the CR transceiver of each SU requires greater computational resources than are required in centralized DSA.

2.3 Cognitive Radio

CR is a paradigm that enables an xG network to use spectrum in a dynamic manner, based on the concept of DSA. Through DSA, frequency spectrum can be shared among PUs and SUs in a dynamically changing radio environment. CR and DSA techniques can be integrated into traditional wireless communications systems to achieve better flexibility of radio resource usage so that the system performance can be improved [8]. For example, load balancing/dynamic channel selection in traditional cellular wireless systems and WLANs, distributed subcarrier allocation in OFDM systems, and transmitter power control in UWB systems can be achieved by using dynamic spectrum access-based CR techniques.

In the early 21st century, Joseph Mitola introduced the idea of SDRs. These radios typically have a radio frequency (RF) front end with a software-controlled tuner [6]. Baseband signals are passed into an analog-to-digital converter. The quantized baseband is then demodulated in a reconfigurable device such as a field-programmable gate array (FPGA), digital signal processor (DSP), or commodity PC. The reconfigurability of the modulation scheme makes it a SDR. In an article published in 2009 [7], Mitola took the SDR concept one step further, coining the term CR [7]. CRs are essentially SDRs with artificial intelligence, capable of sensing and reacting to their environment. Mitola described how a CR could enhance the flexibility of personal wireless services through a new language called the Radio Knowledge Representation

Language (RKRL). Figure 2.3 graphically contrasts traditional radio, software radio, and CR.

The term CR is defined by S. Haykin [8] as an intelligent wireless communication system that is aware of its surrounding environment (i.e., outside world) and uses the methodology of understanding-by-building to learn from the environment and adapt its internal states to statistical variations in the incoming RF stimuli by making corresponding changes in certain operating parameters (e.g., transmitter power, carrier frequency, and modulation strategy) in real time, with two primary objectives, that is, highly reliable communications whenever and wherever needed and efficient utilization of the radio spectrum. FCC explains CR as "A radio that can change its transmitter parameters based on interaction with the environment in which it operates" [9] (see Figure 2.4).

Various definitions of CR can be found in [11–13].

Though more representative of Mitola's original research direction, these interpretations are a bit too futuristic for today's technology. A more common definition restricts the radio's cognition to more practical sensory inputs that

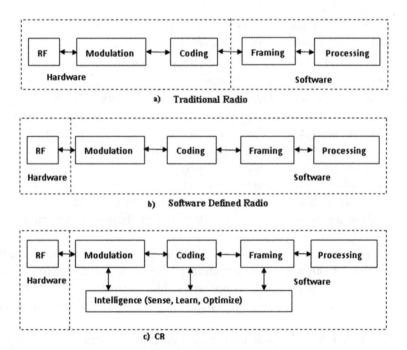

Figure 2.3 Logical diagram of traditional radio, software radio, and cognitive radio.

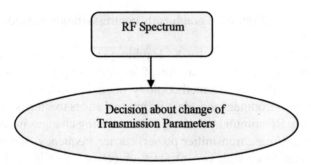

Figure 2.4   Definition of CR according to FCC.

are aligned with typical radio operation. "A radio may be able to sense the current spectral environment, and have some memory of past transmitted and received packets along with their power, bandwidth, and modulation. From all this, it can make better decisions about how to best optimize for some overall goal". Possible goals could include achieving an optimal network capacity, minimizing interference to other signals, or providing robust security or jamming protection.

Capabilities that can be incorporated to allow more efficient and flexible spectrum usage include the following [35]:

- *Frequency Agility*: The ability of CR to change its operating frequency for its adaptation to the environment.
- *Dynamic Frequency Selection*: With this capability, CR senses signals from nearby transmitters to choose an optimal environment to work in.
- *Adaptive Modulation*: By virtue of this ability, CR reconfigured the transmission characteristics and waveforms to exploit all opportunities for spectrum usage in an efficient way.
- *Transmitter Power Control*: CR can allow transmitting at full power limits when necessary but maintain the transmitter to operate at lower levels of transmission power to allow greater sharing of spectrum when higher power operation is not necessary.
- *Location Awareness*: The ability of CR to determine its location and the location of other devices operating in the same spectrum, first to see whether it is permissible to operate at high power levels, and then to select appropriate transmission parameters such as operating frequency and power.
- *Negotiated Use*: The CR may have algorithms enabling the sharing of spectrum in terms of prearranged agreements between a licensee and a

third party. Negotiation can be based on an ad hoc/real-time basis without the need for prior agreements between all the parties.

In order to efficiently utilize the WS, the cognitive cycle [14] is analyzed as shown in Figure 2.5. The cognitive cycle is different from today's handsets that either blast out on the frequency set by the user or blindly take instructions from the network.

According to [14], two main characteristics of the CR are as follows:

- *Cognitive capability*: Cognitive capability refers to the ability of the radio technology to capture or sense the information from its radio environment and to determine appropriate communication parameters and adapt to the dynamic radio environment. The tasks required for adaptive operation in open spectrum is referred to as the **cognitive cycle**. Through this capability, the portions of the spectrum that are unused at a specific time or location can be identified. Consequently, the best spectrum and appropriate operating parameters can be selected.
- *Reconfigurability*: This enables the radio to be dynamically programmed without any modification in hardware components according to the radio environment means; thereby, device is able to change its operating parameters such as operating frequency, modulation, and transmitter power according to information gathered from its radio environment/communication link. This capability is realized through SDR which can tune to any frequency band and receive any modulation over a

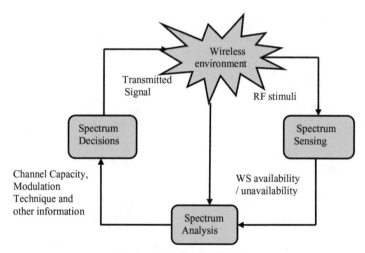

Figure 2.5 Cognitive cycle.

large-frequency spectrum by means of programmable hardware which is controlled by software. Another saying, a SDR can easily switch among multiple wireless protocols or move to different frequencies, waveforms, protocols, or applications, but the user must command it to do so. A complete hardware-based radio system has limited utility since parameters for each of the functional modules are fixed. On the other hand, a SDR extends the utility of the system for a wide range of applications that use different link-layer protocols and modulation/demodulation techniques. Thus, this software-based approach provides reconfigurability to the CR systems. Reconfigurability means to be able to reconfigure the transmission parameters during the transmission. This fact indicates that the CR is capable of configuring both transmitter and receiver parameters in order to switch to different spectrum bands by using appropriate protocols and modulation schemes with assigning appropriate power level of the signal.

Figure 2.6 shows functional components of a more concrete CR architecture. The SDR is accessed via a CR application programming interface (API) that allows the CR engine to configure the radio and sense its environment. The policy-based reasoning engine takes facts from the knowledge base and information from the environment to form judgments about RF spectrum accessing opportunities. In addition to a simple policy-based engine, a learning engine observes the radio's behavior and resulting performance and adjusts facts in the knowledge base used to form judgments.

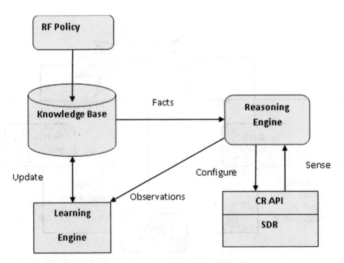

Figure 2.6 Functional portions of a CR, representing reasoning and learning capabilities.

Advantages of CR technology:

- The CR technologies could meet ever-increasing requirements for spectral efficiency, effective etiquette, and resistance to interference.
- This efficiency can also be extended to creating greater device convergence, allowing multiple services to coexist in the same devices, as well as in the same spectrum.
- It is noteworthy that the CR network concept has the potential to explicitly protect the spectrum rights of incumbent license holders.
- CR technology thus empowers radios to observe more flexible radio etiquettes that were not possible in the past. CR offers hope to meet increasing demand of spectrum with a system that is compatible with existing deployed wireless systems, stimulates new innovation, reduces regulatory burden, encourages market competition, preserves the rights of incumbent spectrum license holders, and benefits the populace overall [14].

2.4 Spectrum Sensing Challenges

CR seems conceptually very simple, but its realization is highly challenging. Some of challenges in spectrum sensing are described below.

2.4.1 Hidden Primary User Problem

In CR network, primary receiver may not be cooperating with secondary receiver, and hence, the CR may not be able to avoid generating interference to the PUs when the primary transmitter is out of the CR's detectable range. This problem is referred to as the **hidden node problem or hidden PU problem**. The hidden PU problem is caused by many factors including severe multipath fading or shadowing observed by secondary users while scanning for PUs transmission [15–18]. Figure 2.7 shows an illustration of a hidden node problem where the circle shows the operating range of the PU. A CR user cannot distinguish between a deeply faded signal and an idle one. To address these issues, cooperative spectrum sensing is used [19, 20].

2.4.2 Channel Uncertainty

Under fading or shadowing channel, a low received signal strength does not necessarily imply that the PU is located out of the SU's interference range, as the primary signal may be experiencing a deep fade or being heavily shadowed by obstacles. Therefore, spectrum sensing is challenged by such channel

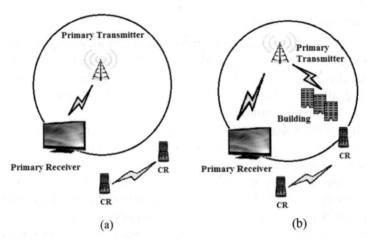

Figure 2.7   Hidden node problem (a) primary receiver without shadowing effect, (b) primary receiver signal with shadowing effect.

uncertainty [21]. Various aspects of cooperative sensing can be utilized to mitigate the effect of shadowing, channel uncertainty, and the primary receiver location uncertainty [22].

2.4.3 Noise Uncertainty

Detector sensitivity is affected by the noise power [21–23]. A priori knowledge of noise power, however, is not available in practice. Unfortunately in practice, the uncertainty always exists due to uncertainty of the measurements, changes in thermal noise caused by temperature variations and of the time variation of noise power. Spectrum sensing is further challenged by noise uncertainty when energy detection is used as the underlying sensing technique.

2.4.4 Cross-Layer Design

Since the CR is capable of adjusting its radio parameters such as physical link quality, radio interference, radio node density, network topology modulation technique, frequency, power, transmitter bandwidth, antenna, modulation and coding techniques, or traffic demand [24, 25] dynamically to choose the best available option [26], CRs require an advanced control and management framework with support for cross-layer information. Spectrum handoff and mobility management are some new challenges which are required to do a cross-layer design, in terms of quality-of-service. This requires a more robust, efficient, and reconfigurable hardware and software architecture. But the CR receivers are capable of processing the narrowband baseband signals with reasonably low-complexity and low-power processors. But in order to not to miss any spectrum opportunities, CR should be capable to capture and analyze a relatively larger band which demands for high-speed signal processors (DSPs, FPGAs) with low delay, high sampling rate, and high-resolution analog-to-digital converters (ADCs) with large dynamic range [27]. The large operating bandwidths impose additional requirements on antennas and power amplifiers. If this hardware requirement criterion cannot be fulfilled, the spectrum sensing will not be implemented. In [28, 29], two different architectures for sensing in CR are given:

- Single-radio: In this type, only a specific time slot is allocated for spectrum sensing. Some portion of the time slot is utilized for spectrum sensing rather for data transmission, and hence, spectrum opportunity is insufficiently utilized. As a result of this limited sensing duration, only certain accuracy can be guaranteed for spectrum sensing results.

Advantage of single-radio is simple and cost-effective.

- Dual-radio: One radio channel is dedicated for continuous spectrum monitoring, and other radio chain is used for data transmission and reception. This approach requires more power consumption as continuous spectrum monitoring is done.

GNU Radio [29], Universal Software Radio Peripheral (USRP) [30], Shared Spectrum's XG Radio [31], and Universal Software Radio Peripheral 2 (USRP2) [32] are the available hardware and software platforms for the CR.

2.4.5 Spread Spectrum Primary Users Detection

Detection of spread spectrum PUs is difficult as the power of the primary user is distributed over a wide frequency range even though the actual information bandwidth is much narrower [33]. This problem can be partially avoided if the hopping pattern is known and perfect synchronization to the signal can be achieved. However, it is complicated to design algorithms that can do the estimation in code dimension.

2.4.6 Sensing Duration and Frequency

The CR should require the band immediately when SU senses the PUs which impose the constraints on how much time is used for monitoring the spectrum use versus actual data transmission and how often spectrum must be sensed. Selection of sensing parameters brings about a trade-off between the speed (sensing time) and reliability of sensing. The optimum value of sensing frequency depends on the capabilities of CR itself and the temporal characteristics of primary users in the environment. Sensing time must be very small for the public safety spectrum in order to avoid interference, while less frequent sensing may be allowed for the TV spectrum where the spectrum usage varies over a much larger timescale [34]. In the IEEE 302.22 draft standard, the sensing period is selected as 30 s. The channel detection time, channel move time, and some other timing-related parameters are also defined in the standard [34].

Since it is not possible to transmit on a licensed band and sense it simultaneously, sensing has to be interleaved with data transmission. The various methods are studied in [35] for obtaining detection time using numerical optimization, for maximizing channel efficiency for a given detection probability. The sensing can be performed without losing useful bandwidth in orthogonal frequency division multiplexing (OFDM) where guard band between OFDM symbols is replaced by quiet periods and sensing is performed during these quiet periods. Sensing time can be decreased by sensing only changing parts of the spectrum instead of the entire target spectrum. Direct frequency hopping (DFH) method can be applied based on the assumption of having more than a single channel for CR [36]. During operation on a working channel, the intended channel is sensed in parallel. If there is an available channel, channel switching takes place and one of the intended channels becomes the working channel. The access point (AP) decides the channel-hopping pattern and broadcasts this information to connected stations [37].

2.4.7 Decision Fusion in Cooperative Sensing

In the case of cooperative sensing, sharing information among CRs and combining results from various measurements is a challenging task. The shared information can be soft or hard decisions made by each cognitive device. The results presented in [37–39] show that soft information-combining outperforms hard information-combining method in terms of the probability of missed opportunity. On the other hand, hard decisions are found to perform as good as soft decisions when the number of cooperating users is high [40].

2.4.8 Security and Trusted Access

In CR networks, wireless devices do not operate in fixed spectrums, but search and find appropriate spectrums to operate in, so some security issues have to be addressed. At first, CR networks inherit security problems from general wireless networks. Secondly, CR networks have some new security problems, such as spectrum misusage and selfish misbehaviors, primary user emulation (PUE) attack [41, 42], and eavesdropping. Military and intelligence wireless systems require application-specific secure wireless communication. How to secure a CR network by understanding identity, earning and using trust for individual devices, and extending the usage of trust to networking has been discussed in [43].

2.4.9 Spectrum Sensing in Multidimensional Environment

In the CR network, there will be increased possibility of multiple CR networks operating over the same licensed band. As a result, spectrum sensing will be complicated by uncertainty in aggregate interference (e.g., due to the unknown number of secondary systems and their locations). The aggregate interference may turn out to be harmful even though a primary system may be out of any secondary system's interference range. This uncertainty calls for more sensitive detectors as a secondary system may harmfully interfere with primary systems located beyond its interference range, and hence should be able to detect them [21]. To overcome the problem, a few aspects of cooperation in spectrum sensing must be considered such as the distributed information (transmitted power, frequency of other users, location, etc.), the way to cooperate with other SUs, and the need for PUs involving in the cooperation [22, 41, 44].

2.4.10 Interference Temperature Measurement

Two primary challenges in interference temperature concept are [45] the determination of the background interference environment and the measurement of the interference temperature. Moreover, current interference model does not consider multi-user.

2.4.11 Complexity Issue

CR is being proposed as a future way of tackling the problem of increasingly radio spectrum. To achieve this, it requires that the communications nodes themselves are intelligently capable of sensing, and dynamically selecting, the appropriate spectral resources without causing excessive interference on other users. To achieve this, researchers are proposing a variety of increasingly complex methods of implementing CR, which incorporate SDR, dynamic spectrum management, and intelligence [46, 47]. The drawback of this complexity is that it is predicted that CR is still years away from implementation. The challenge is to understand whether such complexity is justified, and what benefits it brings to overcome the current regulatory-constrained spectral assignment process. It is foreseen that it should be possible to develop reduced complexity strategies that will deliver much of the functionality of proposed systems, enabling more rapid adoption, and wider use in systems where CR is currently not being considered due to prohibitive complexity.

2.5 Characteristics of Spectrum Sensing

Spectrum sensing is based on a well-known technique called signal detection. Signal detection can be described as a method for identifying the presence of a signal in a noisy environment. Signal detection has been thoroughly studied for radar purposes since the fifties [48]. Analytically, signal detection can be reduced to a simple identification problem, formalized as a hypothesis test [49]. The Neyman–Pearson criteria say that the spectrum detection is a binary hypothesis sensing problem given as:

$$
\begin{aligned}
H_0 &: y(n) = \eta(n) & n &= 1, 2, 3, \ldots \\
H_1 &: y(n) = s(n)\, h + \eta(n) & n &= 1, 2, 3,
\end{aligned}
\tag{2.1}
$$

where $s(n)$ indicates the PU, h is the channel coefficient, and $\eta(n)$ is the additive white Gaussian noise which is assumed to be independent and identically distributed random variable with zero mean and variance σ_η^2. Here, the hypothesis H_0 means the absence of signal and H_1 its presence. We can define four possible cases for the detected signal:

1. Declaring H_1 under H_1 hypothesis which leads to *Probability of Detection* (P_d)
2. Declaring H_0 under H_1 hypothesis which leads to *Probability of Missing* (P_m)
3. Declaring H_1 under H_0 hypothesis which leads to *Probability of False Alarm* (P_{fa})
4. Declaring H_0 under H_0 hypothesis.

Performance of spectrum sensing algorithm is indicated by two metrics: by a probability of detection, P_d, which is the probability of the algorithm correctly detecting the presence of the PU; and by a probability of false alarm, P_{fa}, which defines probability of the algorithm mistakenly declaring the presence of PU. Thus, false alarm error leads to inefficient usage of the available spectrum opportunities. If H_0 is decided under H_1 hypothesis, then it leads to probability of missing, P_m, that is probability of deciding that there's no PU signal while PU signal actually exists. Another saying, it is the probability of signal missing. Literally, the aim of the signal detector is to achieve correct detection all of the time, but this can never be perfectly achieved in practice because of the statistical nature of the problem. Therefore, signal detectors are designed to operate within prescribed minimum error levels. Missed detections are the biggest issue for spectrum sensing, as it means possibly interfering with the primary system. Nevertheless, it is desirable to keep false alarm probability as low as possible, so that the system can exploit all possible transmission opportunities.

In a wireless communication, it is reasonable to assume that the spectrum sensing detecting device does not know the location of the transmitter, two cases:

- A low h is solely due to the path loss (distance) between the transmitter and the sensing device, meaning that the latter is out of range and can safely transmit.
- A low h is due to shadowing or multipath, meaning that the sensing device might be within the range of the transmitter and can cause harmful interference.

2.6 Spectrum Sensing Methods

The spectrum sensing method can be categorized as follows based on whether they require knowledge of signal power, noise power, or both (see Figure 2.8):

- Methods requiring both signal and noise power information: likelihood ratio test, cyclostationary detection, and matched filter.
- Methods requiring only noise power/semi-blind detection: energy detection, wavelet-based method.
- Methods requiring no information on signal source or noise power/total blind detection: eigenvalue-based sensing; covariance-based and blindly combined energy detection.

The various spectrum sensing methods including likelihood ratio test (LRT), energy detection (ED) method, matched filtering (MF)-based method, cyclostationary detection method, multitaper spectrum estimation, and covariance-based, wavelet transform-based estimation, and support vector machine (SVM) have been studied for the detection of PUs. Each of this method has different requirements together with advantages or disadvantages.

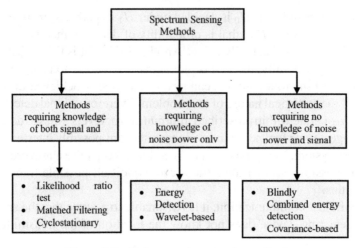

Figure 2.8   Various spectrum sensing methods.

2.6.1 Matched Filtering

Matched filtering is the optimum method for detection of PUs when the transmitted signal is known [49–51]. The main advantage of matched filtering is the short time to achieve a certain probability of false alarm or probability of miss detection as compared to other methods such as cyclostationary. In fact, the required number of samples increases as $O(1/\text{SNR})$ for a target probability of false alarm at low SNRs for matched filtering [26]. However, matched filtering requires CR to demodulate received signals. Hence, it requires perfect knowledge of the primary users' signaling features such as bandwidth, operating frequency, modulation type and order, pulse shaping, and frame format. This is possible only if the PUs cooperate with the SUs or else leads to degradation in the performance of MF-based method [2, 52]. Moreover, since CR needs receivers for all signal types, the implementation complexity of sensing unit is impractically large [53]. Another disadvantage of matched filtering is large power consumption as various receiver algorithms need to be executed for detection.

2.6.2 Cyclostationary Detection

It is a method for detecting PU transmissions by exploiting the cyclostationarity features of the received signals. Instead of power spectral density (PSD), cyclic correlation function is used for detecting signals present in a given spectrum. The cyclostationarity-based detection algorithms can differentiate noise from PUs' signals. This is a result of the fact that noise is wide-sense stationary (WSS) with no correlation, while modulated signals are cyclostationary with spectral correlation due to the redundancy of signal periodicities [54, 55]. Furthermore, cyclostationarity can be used for distinguishing among different types of transmissions and PUs. The cyclostationary detection is used for OFDM-signal detection based on multiple cyclic frequencies [55]. Also in [56, 57], cyclostationary-based detection is applied for OFDM-based digital TV signals for the IEEE 302.22 Wireless Regional Network (WAN) standard. In [58], cyclostationary detection is used when the signal spectrum is partly intercepted; an improved cyclic spectrum estimator as well as detection strategy for unknown feature location is proposed. A new algorithm called cyclostationary feature quickest spectrum sensing algorithm is proposed in [58]. But cyclostationary detection method requires the exact knowledge of the cyclic frequencies of the PUs, which is not possible practically for many spectrum reuse applications [27, 49]. Also this method demands excessive analog-to-digital (A/D) converter requirements and signal processing capabilities, high computational complexity and strength of SCD could be affected by the unknown channel, and sampling time error and frequency offset could affect the cyclic frequencies and need to know the cyclic frequencies of the primary signal which may not be realistic for many spectrum-reuse applications.

2.6.3 GLRT

Likelihood ratio test (LRT) is proven to be optimal, but it is very difficult to use, as it requires exact channel information and distributions of the source signal and noise which are difficult to obtain practically [49]. An iterative GLRT sensing algorithm and a simple non-iterative GLRT sensing algorithm are developed in [59] for slow- and fast-fading channels. It is also used for spectrum sensing in OFDM access (OFDMA) systems and in multiple-input multiple-output (MIMO) systems. The non-iterative GLRT sensing algorithm offers the best performance under slow-fading channels, fast-fading channels, OFDMA systems, and MIMO systems, and it significantly outperforms several spectrum sensing methods when there is noise uncertainty.

2.6.4 Multitaper Spectrum Estimation

Multitaper spectrum estimation is shown to be an approximation to maximum likelihood PSD estimator, and it is nearly optimal for wide-band signals

[50, 51]. Although the complexity of this method is smaller than the maximum likelihood estimator, it is still computationally demanding.

2.6.5 Wavelets

Wavelets [62] are used for detecting edges in the PSD of a wideband channel. Once the edges, which correspond to transitions from an occupied band to an unoccupied (vacant/empty band) or vice versa, are detected, the powers within bands between two edges are estimated. Using this information and edge positions, the frequency spectrum can be identified as vacant or occupied. Fast sensing is possible by focusing on the frequencies with active transmissions after an initial rough scanning. A testbed implementation of this algorithm is explained in [63].

2.6.6 Energy Detection

Energy detection (ED) method is a semi-blind detection and requires knowledge of noise power only for signal detection. In [64], sets of receiver operating characteristic (ROC) curves are drawn for several time–bandwidth products for spectrum sensing using energy detection. In [49], blindly combined energy detection method, which uses the spatial correlation of received signals based on energy detection, is analyzed for independent and identically distributed (i.i.d.) source signal which is frequency-modulated wireless signal operated in vacant TV channels with a bandwidth less than 200 kHz, the flat-fading and the multipath fading channel. The experimental study for spectrum sensing in 2.4 GHz ISM band over 35 MHz of bandwidth using energy detection for sine wave carrier and QPSK sensing is evaluated in [64]. In the same reference, minimum detectable signal levels set by the receiver noise uncertainties are measured.

2.6.7 Covariance-based Method

In [49, 65, 66], a new spectrum sensing algorithm based on covariance of the received signal called covariance absolute value (CAV) is proposed. CAV is a blind detection method and uses space–time signal correlation for signal detection which does not require any knowledge of noise and signal power. The covariances of signal and noise are generally different, which can be used in detection of PU. The performance of ED and CAV is evaluated for Advanced Television Systems Committee (ATSC) (in USA and South Korea) having 6 MHz bandwidth. But CAV method is very sensitive to the signal correlation.

2.6.8 Other Spectrum Sensing Methods

Random Hough transformation of received signal is used in [67] for identifying the presence of radar pulses in the operating channels of IEEE 302.11 systems. This method can be used to detect any type of signal with a periodic pattern as well. In [68], a support vector machine (SVM), which is a data mining method, is applied for detection of PU signal. It shows superior performance compared to energy detection. In [69], hybrid architecture, associating energy and cyclostationary detectors for spectrum sensing that improves the ability of conventional energy detector to detect the PU in the presence of noise uncertainty. In a constant noise environment, the performance of the proposed detector approaches that of an ideal radiometer. In [70], spectrum sensing method based on higher-order statistics is discussed. Here, bispectrum estimation method for its good performance in signal processing field is used.

2.7 Analysis of Energy Detection and Covariance Absolute Value Method

2.7.1 Energy Detection

Energy detection (ED) method compares the energy of the received waveform over a specified time with a threshold obtained for a given P_{fa}, limit to decide whether primary signal exists or not. Note that for a given signal bandwidth B, a pre-filter matched to the bandwidth of the signal needs to be applied as shown in Figure 2.9.

The decision statistic for energy detector is [22, 49, 71]:

$$T_{ED} = \frac{1}{N} \sum_{n=0}^{N-1} (|y(n)|)^2 \qquad (2.2)$$

where N is the total number of taken signal samples. This decision statistic is compared with a threshold β_{ED}. If $T_{ED} > \beta_{ED}$, the signal is assumed to be present; otherwise, it is assumed to be absent. In case of large N, according to

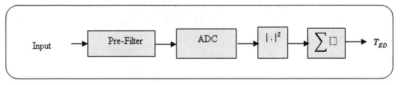

Figure 2.9 Energy detection.

central limit theorem (CLT), the probability distribution function of T_{ED} can be approximated by Gaussian distribution. The expressions for probability of detection and false alarm can be obtained as:

$$P_d = Q\left(\frac{\sqrt{N}\ \left(\beta_{ED} - \left(\sigma_s^2 + \sigma_\eta^2\right)\right)}{\sigma_s^2 + \sigma_\eta^2}\right) \tag{2.3}$$

$$P_{fa} = Q\left(\frac{\sqrt{N}\left(\beta_{ED} - \sigma_\eta^2\right)}{\sigma_\eta^2}\right) \tag{2.4}$$

where σ_s^2 is PU signal variance, σ_η^2 is noise variance, and $Q(t)$ is Q function given by:

$$Q(t) = \frac{1}{\sqrt{2\pi}} \int\limits_{t}^{+\infty} e^{-\frac{u^2}{2}}\, \mathrm{du} \tag{2.5}$$

In energy detection, accurate knowledge of noise power is a must for signal detection. But unfortunately in practice, the uncertainty always exists due to uncertainty of the measurements and of the time variation of noise power. Due to the noise uncertainty [22, 23, 113], the estimated noise power may be different from the actual noise power by a factor γ, called noise uncertainty factor: $\hat{\sigma}_\eta^2 = \gamma\sigma_\eta^2$. The noise uncertainty bound (in dB) is given as:

$$A = \sup\{10\log_{10}\gamma\} \tag{2.6}$$

The noise uncertainty bound is usually 1 to 2 dB. This creates signal-to-noise ratio wall (SNR$_{wall}$) under which it is not possible to detect the signal [21]. The degradation in performance of ED method due to noise uncertainty is studied in detail in [22].

2.7.2 Covariance Absolute Value

CAV is a blind detection method that does not require any knowledge of noise and signal power and uses space–time signal correlation for signal detection. The covariances of signal and noise are generally different. This difference is used in CAV method to differentiate the signal component (PU) from background noise [65, 66].

$$s(l) = [s(N-1-l),\ \ldots,\ s(-l)] \tag{2.7}$$

$$\eta(l) = [\eta(N-1-l),\ \ldots,\ \eta(-l)]$$

$$y(l) = [y(N - 1 - l), \ldots, y(-l)]$$

where N is the total number of signal samples, the range of l is from 0 to $L-1$, the parameter L is called smoothing factor as explained in [65, 66]. $s(l)$ represents signal samples, $\eta(l)$ represents noise samples, and $y(l)$ represents the received signal samples.

$$y(l) = s(l) + \eta(l) \tag{2.8}$$

The autocorrelation of received signal is

$$c(l) = \frac{1}{N} \sum_{n=0}^{N-1} y(n) * y(n-l) \, l = 0, 1, .., L-1 \tag{2.9}$$

The covariance matrix of $y(l)$ under H_0 and H_1 is given as:

$$H_0 \colon R_y(l) = E\left[y(l)y(l)^H\right] = \sigma_\eta^2 I_L \tag{2.10}$$

$$H_1 \colon R_y(l) = E\left[y(l)y(l)^H\right] = R_s(l) + \sigma_\eta^2 I_L$$

where $R_s(l)$ is the covariance matrix of $s(l)$, $\sigma_\eta = E[\eta(l) * \eta(l)^*]$, and I_L is the identity matrix of dimension L. If there is a correlation between the signal samples due to oversampling, the $R_s(l)$ will be different from identity matrix. In the matrix form, the covariance matrix of the received signal is

$$R_y(N) = \begin{bmatrix} c(0) \ c(1) \ .. \ c(L-1) \\ c^*(1) \ c(0) \ .. \ c(L-2) \\ c^*(L-1) \ c^*(L-2) \ .. \ c(0) \end{bmatrix} \tag{2.11}$$

This matrix is Toeplitz Hermitian matrix. Based on this covariance matrix, two metrics T_1 and T_2 are defined as follows:

$$T_1 = c(0) + \frac{2}{L} \sum_{l=1}^{L-1} (L-l) \, |c(l)| \tag{2.12}$$

$$T_2 = c(0) \tag{2.13}$$

The ratio $\frac{T_1}{T_2}$ can be used to detect the presence of the signal. When there is no signal, $\frac{T_1}{T_2} = 1$, and if the signal is present, $\frac{T_1}{T_2} > 1$.

The following decision metric for CAV is derived as:

$$T_{CAV} = \frac{T_1}{T_2} \tag{2.14}$$

According to this decision metric, probability of false alarm and the probability of detection have the following form:

$$P_{fa} = 1 - Q\left(\frac{\frac{1}{\beta_{CAV}}\left(1 + (L-1)\sqrt{\frac{2}{N\pi}}\right) - 1}{\sqrt{2/N}}\right) \tag{2.15}$$

$$P_d = 1 - Q\left(\frac{\frac{1}{\beta_{CAV}} + \frac{\beta_L \ \text{SNR}}{\beta_{CAV} \ (\text{SNR}+1)} - 1}{\sqrt{2/N}}\right) \tag{2.16}$$

where signal-to-noise ratio is given as SNR $= \frac{\sigma_s^2}{\sigma_\eta^2}$, β_{CAV} is the decision threshold, and β_L is a very important parameter for CAV called overall correlation coefficient. It is defined as:

$$\beta_L = \frac{2}{L}\sum_{l=1}^{L-1}(L-l)\,|e(l)| \tag{2.17}$$

where

$$e(l) = \frac{E\left[s(n)s(n-l)\right]}{\sigma_s^2} \tag{2.18}$$

This is the normalized correlation strength between the signal samples (range is between 0 and 1). However, usually, the signal samples should be correlated due to three reasons:

1) The signal is oversampled. Let T_0 be the Nyquist sampling period of signal $s_C\,(t)$ and $s_C\,(nT_0)$ be the sampled signal based on the Nyquist sampling rate. Based on the sampling theorem, signal $s_C\,(t)$ can be expressed as

$$s_C\,(t) = \sum_{n=-\infty}^{\infty} s_C\,(nT_0)g(t - nT_0) \tag{2.19}$$

where $g(t)$ is an interpolation function. Hence, the signal samples $s(n)$ and $s_C\,(nT_s)$ are only related to $s_C\,(nT_0)$. If the sampling rate at the receiver $f_s > 1/T_0$, that is, $T_s < T_0$, then $s(n)$ and $s_C\,(nT_s)$ must be correlated.

2) The propagation channel has time dispersion; thus, the actual signal component at the receiver is given by:

$$s_C(t) = \int_{-\infty}^{\infty} h(\tau) \, s_0 \, (t - \tau) d\tau \qquad (2.20)$$

where $s_0(t)$ is the original transmitted signal, and $h(t)$ is the response of the time-dispersive channel. Since sampling period T_s is usually very small, the integration of Equation (3.20) can be approximated as:

$$s_C(t) \approx T_s \sum_{k=-\infty}^{\infty} h(\mathrm{k}T_s)s_0(t - \mathrm{k}T_s) \qquad (2.21)$$

Hence,

$$s_C(\mathrm{n}T_s) \approx T_s \sum_{k=k_0}^{k_1} h(\mathrm{k}T_s)s_0((n - k)T_s) \qquad (2.22)$$

where $[k_0 \, T_s, \, k_1 \, T_s,]$ is the support of channel response $h(t)$, that is, $h(t) = 0$ for $t \notin [k_0 \, T_s, \, k_1 \, T_s]$. For the time-dispersive channel, $k_0 > k_1,$; thus, the signal samples $s_C \, (\mathrm{n}T_s)$ are correlated, even if the original signal samples $s_0 \, (\mathrm{n}T_s)$ could be i.i.d.

3) The original signal is correlated. In this case, even if the channel is a flat-fading channel and there is no oversampling, the received signal samples are correlated.

Another assumption for the algorithm is that the noise samples are i.i.d. This is usually true if no filtering is used. However, if a narrow-band filter is used at the receiver, the noise samples will sometimes be correlated. To deal with this case, pre-whitening the noise samples (making the samples i.i.d) or pre-transforming the covariance matrix is required.

2.7.2.1 Steps for obtaining the pre-whiten matrix
The received signal is generally passed through a channel selection filter. Let $f(k), k = 0, 1, ..., K$ be the filter. After filtering, the received signal is turned to

$$\tilde{y}(n) = \sum_{k=0}^{k} f(k)y(n - k) \, n = 0, 1, ... \qquad (2.23)$$

$$\tilde{s}(n) = \sum_{k=0}^{k} f(k)s(n - k) \, n = 0, 1, ... \qquad (2.24)$$

$$\widetilde{\eta}(n) \; = \; \sum_{k=0}^{k} f(k)\eta(n-k) \; n \; = \; 0,1,\ldots. \tag{2.25}$$

Then,

$$H_0 : \tilde{y}(n) = \; \widetilde{\eta}(n) \; n \; = \; 1,2,3,\ldots. \tag{2.26}$$
$$H_1 : \tilde{y}(n) = \; \tilde{s}(n) + \widetilde{\eta}(n) \; n \; = \; 1,2,3,\ldots$$

Note that here the noise samples $\widetilde{\eta}(n)$ are correlated. If the sampling rate f_s is larger than the signal bandwidth W, we can down-sample the signal. Let $M \geq 1$ be the down-sampling factor. If the signal to be detected has a narrower bandwidth than W, it is better to choose $M > 1$. For notation simplicity, we still use $\tilde{y}(n)$ to denote the received signal samples after down-sampling, that is, $\tilde{y}(n) \; = \; \tilde{y}(\mathrm{nM})$. Choose a smoothing factor L and define

$$y(n) \; = \; [\tilde{y}(n) \; \tilde{y}(n-1) \ldots \tilde{y}(n-L+1)]^T \; n \; = \; 0,1,\ldots,N-1 \tag{2.27}$$

Define a $Lx(K + 1 + (L-1)M)$ matrix as

$$H \; = \; \begin{bmatrix} f(0) \ldots \ldots f(K) \, 0 \ldots 0 \\ 0 \ldots f(0) \ldots f(K) \ldots 0 \\ \ldots \ldots \ldots \ldots \ldots \ldots \\ 0 \ldots \ldots \ldots f(0) \ldots f(K) \end{bmatrix} \tag{2.28}$$

Let $G = \mathrm{HH}^H$. Decompose the matrix into $= Q^2$, where Q is a L×L Hermitian matrix. The matrix G is not related to signal and noise and can be computed offline. If analog filter or both analog filter and digital filter are used, the matrix G should be revised to include the effects of all the filters. In general, G can be obtained to be the covariance matrix of the received signal, when the input signal is white noise only (this can be done in laboratory offline). The matrixes G and Q are computed only once, and only Q is used in detection. Writing for covariance matrix of R_y under H_1

$$R_y \; = \; R_s + \sigma_\eta^2 G \tag{2.29}$$

where R_s is the statistical covariance matrix of the signal (including fading, multipath, and filtering) and σ_η^2 is the noise variance.

$$\widetilde{R_y} \; = \; Q^{-1} R_y \, Q^{-1} \tag{2.30}$$

Transform the sample covariance matrix to obtain

$$\widetilde{R_s} \; = \; Q^{-1} R_s \, Q^{-1} \tag{2.31}$$

Then,

$$\widetilde{R_y} = \widetilde{R_s} + \sigma_\eta^2 I_L \tag{2.32}$$

If there is no signal, then $\widetilde{R_s} = 0$. Hence, the off-diagonal elements of $\widetilde{R_y}$ are all zeros. If signal presents, $\widetilde{R_s}$ is almost surely not a diagonal matrix. Hence, some of the off-diagonal elements of $\widetilde{R_y}$ should not be zeros. Based on this $\widetilde{R_y}$ given in Equation (2.32), evaluate the Equation (2.10) and Equation (2.11) for deciding T_1 and T_0 which helps in determining the presence of the signal.

2.7.2.2 Flow Chart for CAV method

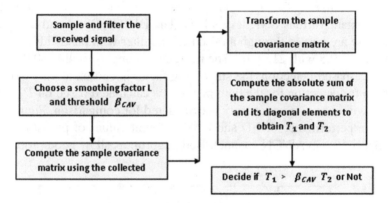

Figure 2.10 Flow chart for CAV method.

2.7.3 Simulation Results for ED without Noise Uncertainty and CAV

To evaluate the performance of energy detector and covariance-based detector methods for IEEE 302.11af, the values for time to vacate band after PUs detection are 2 s with 40% of probability of detection (P_d) and 10% of probability of false alarm (P_{fa}) at signal-to-noise ratio (SNR) level as low as −20 dB along with geo-location accuracy of +/− 50 m [1, 2, 41]. The performance of energy detector and covariance-based detector is analyzed for the PU signal which is TV signal based on DVB–T standard. The DVB-T signal in 2K mode with bandwidth of 8 MHz is generated using MATLAB according to ETSI specifications given in [72] intended for mobile reception of standard TV signal. Table 2.1 gives the specifications for DVB-T in 2K mode.

Table 2.1   DVB-T parameters for 2K mode with 8-MHz channels [72]

Parameter	2K Mode
Number of carriers K	705
Value of carrier number K_{min}	0
Value of carrier number K_{max}	704
Duration T_U	224 μs
Carrier spacing $1/T_U$	4,414 Hz
Spacing between carriers K_{min} and K_{max} $(K-1)/T_U$	7.61 MHz

The received signal is sampled at the rate 36 times the sampling rate at the transmitter. The SNR of the received signal is unknown. In order to use the signals for simulating the algorithms at very low SNRs, we need to add white noise to obtain various SNR levels [71]. The DVB-T signal and the added white noise are passed through a raised cosine filter with bandwidth 8 MHz, rolling factor 0.5 with 217 taps. The number of samples used is 30,000. For covariance-based detection, the smoothing factor is chosen as L = = 6. The threshold for detection for the both methods is calculated based on P_{fa} [73]. The ED without noise uncertainty is considered for comparison. Figure 2.11 shows the performance of ED and CAV different values of probabilities of false alarm P_{fa} in AWGN channel. Both algorithms show good probability

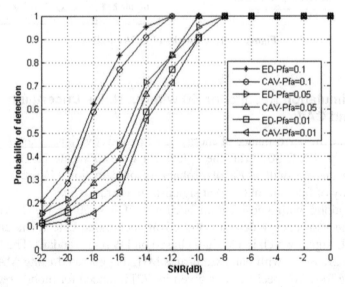

Figure 2.11   Analysis of probability of detection for ED and CAV for different values of P_{fa} with N = 30,000 in AWGN channel.

of detection for $P_{fa} = = 0.1$. For $P_{fa} = 0.1$, ED shows probability of detection for SNR = –20 dB as 0.35 compared to CAV which is 0.28. For $P_{fa} = = 0.05$ and SNR = –20 dB, ED and CAV show probability of detection of 0.21 and 0.16, respectively. For $P_{fa} = = 0.01$ and SNR = –20 dB, ED and CAV show probability of detection of 0.17 and 0.12, respectively. Thus, ED shows the highest detection probability for different values of P_{fa}.

To check the performance of the methods for time-varying channels, the time-varying channel is generated based on the simplified Jakes' model with Doppler frequency DF1 = 100 Hz and DF2 = 1000 Hz ($DF_x = f_d = f_c \frac{v}{c}$, where $DF_x = f_d$ = Doppler shift, v = velocity, c = speed of light, and $f_c = =$ maximum carrier frequency = 302 MHz) for Rayleigh channel with $P_{fa} = = 0.1$ and N = 30,000. DF0 is no Doppler effect. As shown in Figure 2.12, in case of Rayleigh fast time-varying channel at SNR of –18 dB, ED and CAV show probability of detection as 0.20 and 0.15, respectively. For Rayleigh slow time-varying channel, ED and CAV show probability of detection as 0.36 and 0.28, respectively.

For Rayleigh fast time-varying channel at SNR of –14 dB, ED and CAV show probability of detection as 0.58 and 0.52, respectively. For Rayleigh slow time-varying channel at SNR of –14 dB, ED and CAV show probability of

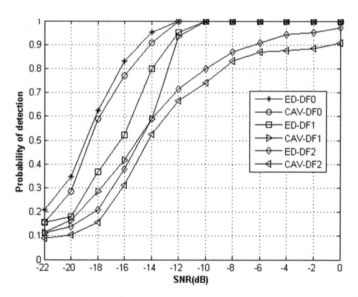

Figure 2.12 Analysis of probability of detection for ED and CAV with $P_{fa} = = 0.1$, N = 30,000 under Rayleigh time-varying channel.

detection as 0.8 and 0.58, respectively. This clearly reveals that the degradation in performance for fast time-varying channel is more compared to slow time-varying channel. The performance of CAV degrades more as compared to ED in Rayleigh fast time-varying channel.

For Rician channel, Doppler frequency of 100 Hz with $P_{fa} = = 0.1$ and $N = 30,000$ is considered, Figures 2.13 and 2.14 reflect its effect on the probability of detection for ED and CAV for $K = 1$ and $K = 10$. For $K = 1$ and SNR = –14 dB, ED and CAV show probability of detection as 0.62 and 0.52, respectively. For $K = 10$ and SNR = –14 dB, ED and CAV show probability of detection as 0.33 and 0.58, respectively. The performance of both methods degrades, but this degradation is more with $K = 1$ compared to $K = 10$.

The effect of time-varying channels at Doppler frequency of 100 Hz on the receiver operating characteristic (ROC) curve is analyzed for both the methods by plotting Figures 2.15 and 2.16 with values SNR = –14 dB, $N = 30,000$, and $P_{fa} = = 0.1$. Figure 2.15 reflects ROC curve under Rayleigh channel, and there is performance deterioration for both ED and CAV, but ED provides probability of detection better than CAV.

Figure 2.16 reflects ROC curve under Rician channel with $K = 1$, and there is performance deterioration for both ED and CAV at SNR of –14 dB compared to $K = 10$. But if we compare the curves with $K = 1$ and $K = 10$, the performance

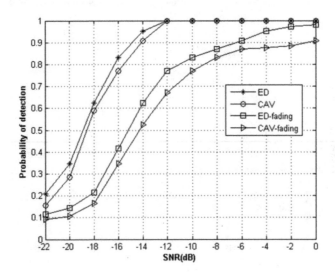

Figure 2.13 Analysis of probability of detection for ED and CAV for $P_{fa} = 0.1$ and $N = 30,000$ under Rician time-varying channel with $K = 1$.

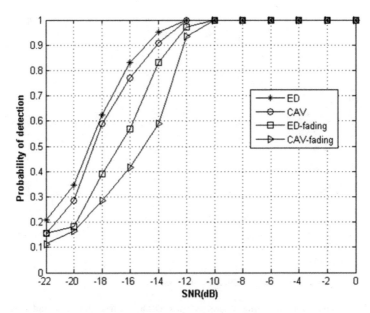

Figure 2.14 Analysis of probability of detection for ED and CAV for $P_{fa} = 0.1$ with $N = 30,000$ under Rician time-varying channel with $K = 10$.

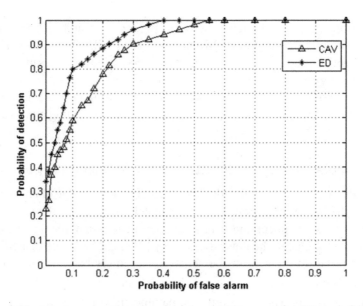

Figure 2.15 Receiver operating characteristics (ROC) curve for ED and CAV with $P_{fa} = = 0.1$, SNR $= -14$ dB and $N = 30,000$ under Rayleigh time-varying channel.

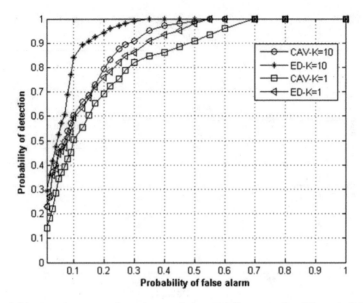

Figure 2.16   Receiver operating characteristics (ROC) curve for ED and CAV with $P_{fa} = 0.1$, SNR $= -14$ dB and $N = 30{,}000$ under Rician time-varying channel for $K = 1$ and $K = 10$.

of detection schemes is better for $K = 10$. Figures 2.15 and 2.16 reveal that ROC curve will be poor at low SNR values than -14 dB. The performance of energy detection and covariance-based spectrum sensing degrades in fast time-varying fading channels. This degradation in performance occurs less in case of slow time-varying channels. The energy detection without noise uncertainty outperformed the covariance-based method in case of fading time-varying channels.

2.7.4 Simulation Results for ED with Noise Uncertainty and CAV

As already mentioned in energy detection, accurate knowledge of noise power is a must for signal detection. But unfortunately in practice, the uncertainty always exists due to uncertainty of the measurements and of the time variation of noise power. Due to the noise uncertainty [21, 22, 65], the estimated noise power may be different from the actual noise power by a factor γ, called noise uncertainty factor: $\widehat{\sigma}_\eta^2 = \gamma \sigma_\eta^2$. The noise uncertainty bound is given by Equation (2.6). The various simulation parameters will be taken same as in Section 2.7.3.

The number of samples used is 30,000. For covariance-based detection, the smoothing factor is chosen as L = 6. The threshold for detection for the both methods is calculated based on P_{fa} [74]. Figure 2.17 shows the performance of ED and CAV for probabilities of false alarm P_{fa} = 0.1 in AWGN channel. "ED-x dB" means energy detection with x-dB noise uncertainties. ED and CAV show probability of detection for SNR = −20 dB as 0.35 and 0.28, respectively. For noise uncertainty such as ED–0.4 dB, ED–1.1 dB, ED–1.6 dB, and ED–1.8 dB, probability of detection is 0.247, 0.239, 0.238, and 0.237, respectively [74]. Comparison of the curves in Figure 2.17 for all values of SNR clearly indicates that even if the noise uncertainty is only 0.4 dB, the probability of detection of energy detection is much worse than CAV. On the other hand, Figure 2.16 also shows that ED with the exact knowledge of noise power performs better compared to CAV method.

To check the performance of the methods for time-varying channels, the time-varying channel is generated based on the simplified Jakes' model with Doppler frequency DF1 = 100 Hz and DF2 = 1000 Hz for Rayleigh channel with P_{fa} = = 0.1 and N = 30,000. As shown in Figure 2.17, in case of Rayleigh slow time-varying channel at SNR of −14 dB, ED, CAV, ED–0.4 dB,

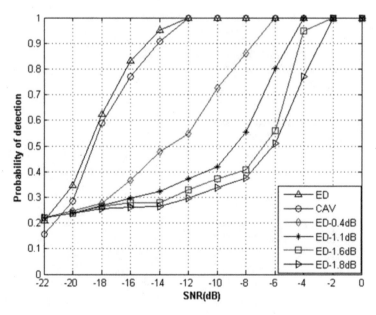

Figure 2.17 Probability of detection versus SNR for P_{fa} = 0.1 with N = 30,000 in AWGN channel.

and ED–1.1 dB show probability of detection as 0.8, 0.58, 0.21, and 0.18, respectively. For Rayleigh fast-fading channel at SNR of –14 dB, ED, CAV, ED–0.4 dB, and ED–1.1 dB show probability of detection 0.6, 0.52, 0.15, and 0.14, respectively. If the curves as shown in Figure 2.18 are compared for slow and fast time-varying channels for all values of SNR, CAV shows good probability of detection compared to ED with noise uncertainty. But the probability of detection for CAV degrades in fast time-varying channel compared to slow time-varying channel. Same is true for ED.

For Rician channel, Doppler frequency of 100 Hz with $P_{fa} = 0.1$ and $N = 30,000$ is considered, and Figure 2.19 reflects its effect on the probability of detection for ED and CAV for $K = 1$ and $K = 10$. At $K = 10$ and SNR = –14 dB, ED, CAV, ED–0.4 dB, and ED–1.1 dB show probability of detection as 0.33, 0.6, 0.21, and 0.18, respectively. For $K = 1$ and SNR = –14 dB, ED, CAV, ED–0.4 dB, and ED–1.1 dB show probability of detection as 0.62, 0.52, 0.15, and 0.14, respectively. By comparing the curves for all values of SNR with $K = 1$ and $K = 10$, it reveals that CAV performs well compared to ED with noise uncertainty. The probability of detection for Rician time-varying channel with $K = 10$ is good, compared to $K = 1$.

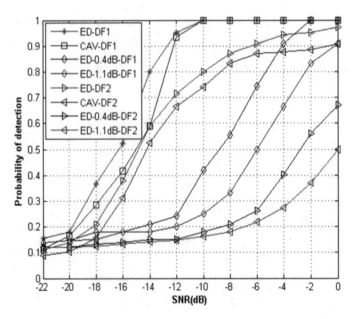

Figure 2.18   Probability of detection versus SNR for $P_{fa} = = 0.1$, $N = 30,000$ under Rayleigh time-varying channel.

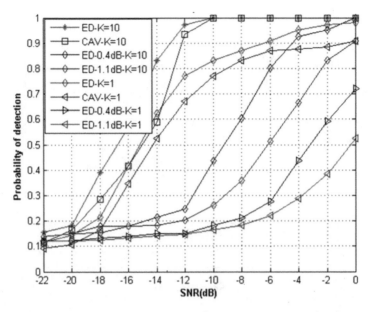

Figure 2.19 Probability of detection versus SNR for $P_{fa} = 0.1$ and $N = 30,000$ under Rician time-varying channel with $K = 10$ and $K = 1$.

To test the impact of smoothing factor, fix $P_{fa} = 0.1$, SNR = –20dB, and $N = 30000$ and vary smoothing factor L from 5 to 10. Figure 2.20 shows the results for probability of detection. A shown in Figure 2.20, since ED is not dependent on L, there is a line parallel to x-axis. For CAV, as L increases from 5 to 9, the probability of detection increases from 0.22 to 0.25. From the graph, it reveals that probability of detection is not very sensitive for smoothing factor greater than 9. Smaller the value of L is the lower the complexity in practice. However, it is difficult to choose the best smoothing factor so that computational complexity reduces as it is related to the signal property which is unknown.-

For testing the impact of overall correlation coefficient on probability of detection for CAV, fix $P_{fa} = = 0.1$, SNR = –14 dB, and $N = 30,000$. As the overall correlation coefficient increases, the probability of detection also increases. Figure 2.21 reveals this fact. The theoretical probability of detection for CAV is also plotted. Further, the probability of detection for ED and ED–0.4 dB is plotted. It clearly reveals that ED and ED–0.4 dB are not dependent on correlation coefficient, so there is a line parallel to x-axis. ED with 0.4 dB noise uncertainty shows poor probability of detection.

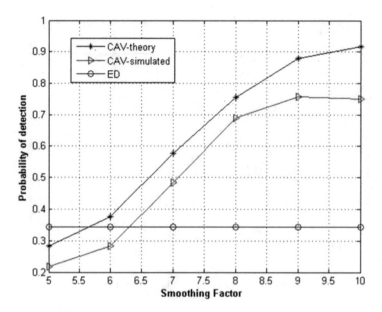

Figure 2.20   Probability of detection versus smoothing factor for $P_{fa} = 0.1$, SNR = –20 dB, and $N = = 30,000$.

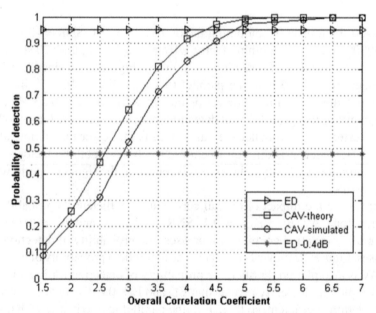

Figure 2.21   Probability of detection versus overall correlation coefficient for $P_{fa} = 0.1$, SNR = –14 dB, and $N = 30,000$.

The simulation results in Section 2.6.4 show that energy detection with accurate noise power can be applied for signal detection and outperformed the covariance method. It is shown that covariance-based detection outperformed energy detection with noise uncertainty present in case of fast and slow time-varying channels. The performance of covariance-based detection degrades in fast time-varying compared to slow time-varying fading channels. The probability of detection for covariance-based detection improves as smoothing factor and overall correlation coefficient increase.

2.8 Hybrid Detection Method

In this section, the need for hybrid detection method for spectrum sensing is analyzed by comparing ED and CAV methods [76, 78].

2.8.1 Comparison between ED- and CAV-based Detection

Number of samples required to achieve P_d and $P_{fa} > 0$

$$N_{ED} = \frac{2\left(Q^{-1}(P_{fa}) - Q^{-1}(P_d)\right)^2}{\text{SNR}^2} \qquad (2.33)$$

$$N_{CAV} = \frac{2\left(Q^{-1}(P_{fa}) - Q^{-1}(P_d) + \frac{L-1}{\sqrt{\pi}}\right)^2}{(\beta)_L\,(\text{SNR})^2} \qquad (2.34)$$

Equation 3.34 indicates that for a fixed P_d and P_{fa}, N_{CAV} is only dependent on smoothing factor and overall correlation strength. Comparing Equations (2.33) and (2.34), if $N_{CAV} < N_{ED}$ is needed, then

$$\beta_L > 1 + \frac{L-1}{\sqrt{\pi}(Q^{-1}(P_{fa}) - Q^{-1}(P_d))} \qquad (2.35)$$

This holds true if signal samples are highly correlated [65, 66].

Computational complexity

ED requires N multiplications and additions, and CAV requires L times that of ED computations.

ED method is one of the simplest methods of spectrum sensing. The energy detection performance is very susceptible to changing noise level as shown in Section 2.6.4. This method strongly depends on noise power; practically, it is difficult to obtain the accurate noise power. In the case of noise uncertainty as shown in [21, 22], there exists a minimum SNR level below which no detection is possible called SNR_{wall}. CAV method has its own limitations. It is highly sensitive to input signal correlation. Overall correlation coefficient (β_L) as defined in Equation (2.21) is a good measure of signal correlation. As signal correlation increases, as shown in Figure 2.19 in Section 2.6.4, the probability of detection also increases; otherwise, its performance degrades.

Using Equations (2.3), (2.4), (2.15), and (2.16), the variation of P_d with β_L for ED and CAV detection methods is plotted [76, 77].

In Figure 2.22, P_d vs. β_L for CAV and ED algorithms for different values of SNR is plotted. In Figure 2.23, P_d vs. β_L for CAV and ED algorithms for different values of number of samples N is analyzed. In Figure 2.24, P_d vs. β_L for CAV and ED algorithms for different values of P_{fa} is traced. In Figure 2.25, P_d vs. P_{fa} for CAV and ED algorithms for different values of β_L is drawn.

Figures 2.22–2.24 reveal that since ED does not dependent on signal correlation strength β_L, a line parallel to x-axis is obtained in the all the figures. From the figures, it is clear that P_d (ED) – P_d (CAV) is a monotonically

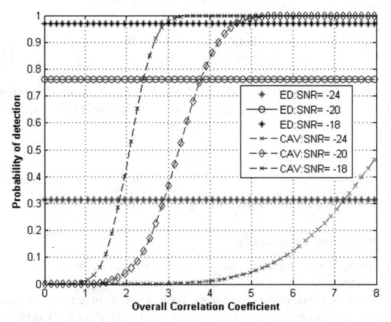

Figure 2.22   Analysis of ED and CAV with N = 40,000, P_{fa} = 0.1, and L = 9 for different values of SNR.

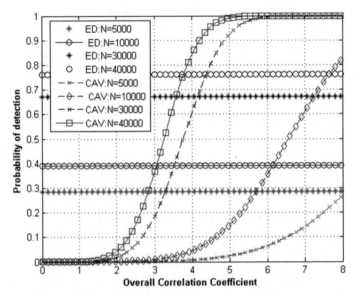

Figure 2.23 Analysis of ED and CAV with SNR = –20 dB, $P_{fa} = 0.1$, and $L = 9$ for different values of N.

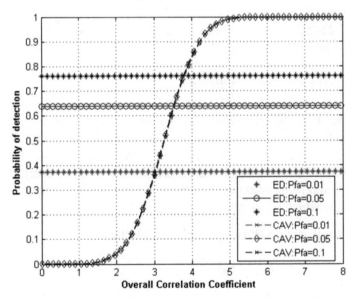

Figure 2.24 Analysis of ED and CAV with SNR = –20 dB, $N = 40,000$, and $L = 9$ for different values of P_{fa}.

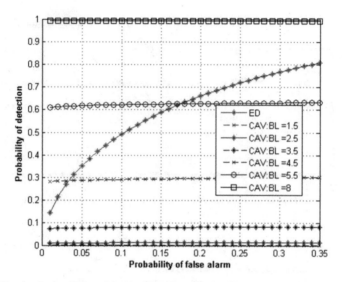

Figure 2.25   Analysis of ED and CAV with SNR $=-20$ dB, $N = 40,000$, and $L = 9$ for different values of β_L.

decreasing function of β_L. Also for a given value of N, L, SNR, and P_{fa}, there is a specific value of β_L below which energy detection is more effective than CAV and above which CAV is more effective than ED. Further, Figures 2.22–2.24 denote that this crossing point decreases with increase in SNR, N, and P_{fa}, respectively.

2.8.2 Hybrid Detection Method

It is clear from Figures 2.21–2.24 that CAV underperforms in low correlation, so the effective measures should be applied to take the advantage of both the methods based on the signal correlation strength. For spectrum sensing, the ED can be used in low correlation and CAV can be used in case of high correlation. In order to detect signal, the decision threshold for hybrid detection method should be derived based on the incoming signal correlation strength for selecting which method to apply.

2.8.2.1 Analysis of Hybrid Detection Method

Define

$$s(l) = [s(N-1-l), \ldots, s(-l)]$$
$$\eta(l) = [\eta(N-1-l), \ldots, \eta(-l)] \tag{2.36}$$
$$y(l) = [y(N-1-l), \ldots, y(-l)]$$

where $s(l)$ represents signal samples and $\eta(l)$ represents noise samples.

$$y(l) = s(l) + \eta(l) \tag{2.37}$$

The sample autocorrelation coefficient $(c(l))$ can be obtained as

$$c(l) = \frac{1}{N} y(0)^T * y(l) \qquad l = 0, 1, 2, \ldots, L-1 \tag{2.38}$$

The decision statistic for hybrid detection method is defined as

$$\beta_{HD} = \frac{2}{L} \sum_{l=1}^{L-1} (L-l) \; |c(l)| \tag{2.39}$$

The mean and variance of $|c(l)|$ can be obtained as shown in [65, 66], which gives the β_{HD} as an unbiased estimator of β_L.

2.8.2.2 Probability distribution function of β_{HD}

The central limit theorem (CLT) for estimation of pdf of β_{HD} cannot be applied as the small value of smoothing factor L is considered (approximately O(10)), but its pdf can be calculated assuming the autocorrelations of received signal $c(l)$'s to be independent of each other. By using theory of folded normal distribution [76–77]. The pdf of a folded normal distribution is the addition of two normal for positive values, and 0 otherwise; thus, pdf of $|c(l)|$ is

$$\text{pdf}(|c(l)|) = N(\gamma_l, \alpha_l) + N(-\gamma_l, \alpha_l) \tag{2.40}$$

where

$$\gamma_l = E(c(l)), \alpha_l = \text{Var}(c(l))$$

Hence, using Equation (2.40), pdf of β_{HD} is

$$\text{pdf}(\beta_{HD}) = \frac{L}{2} (\text{pdf}(c(1)) * \text{pdf}(c(2)) \ldots * \text{pdf}(c(L-1)) \tag{2.41}$$

2.8.2.3 Threshold calculation for hybrid detection method

Threshold (β_{HD}) for deciding when to use ED or CAV method can be obtained by comparing Equations (2.3) and (2.16).

$$p_d \, (\text{ED}) = p_d \, (\text{CAV})|_{\beta_{HD}}$$

$$\beta_{HD} = \frac{\beta_{CAV}(\text{SNR}+1)}{\text{SNR}} \left(Q^{-1}(1 - (p_d \, (\text{ED})) \sqrt{\frac{2}{N}} + 1 - \frac{1}{\beta_{CAV}} \right) \tag{2.42}$$

2.8.2.4 Computational Complexity

If $\beta_L \leq \beta_{HD}$, ED is selected or else CAV method is selected; hence, the computational complexity for hybrid detection method will be N multiplications and additions, or L times N multiplications and additions.

Design steps for hybrid detection method (see Figure 2.26)

Step 1: Calculate the estimate of β_L from the signal samples.
Step 2: Compare the estimate with the threshold.
Step 3: Choose ED or CAV depending on the result of step 2.

2.8.3 Simulation Results for Hybrid Detection Method

In Figure 2.27, the performance of hybrid detection method is noted by plotting the P_d vs β_L for different values of number of samples N with $L = 9$, SNR = -20 dB, and $P_{fa} = 0.01$, and this is compared with CAV and ED methods. It is noted that hybrid detection method is performing better in low correlation strength compared to CAV and ED in case of high β_L. The behavior near β_{HD} can be clarified by using Equation (2.42) as follows: Depending on the incoming signal strength, sometimes ED is selected and sometimes CAV is selected based on $\beta_L \leq \beta_{HD}$ and $\beta_L \geq \beta_{HD}$, respectively.

Further, P_d vs. β_L with SNR = -20 dB, N = 40,000, P_{fa} = 0.01, and L = 9 for different values of P_{fa} is plotted as shown in Figure 2.28.

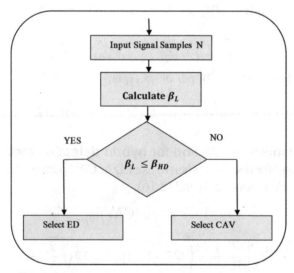

Figure 2.26   Flow Chart for Hybrid Detection Method.

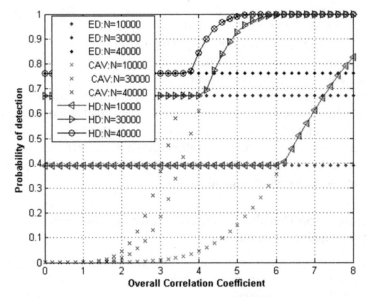

Figure 2.27 Analysis of hybrid detection, ED and CAV method with SNR = –20 dB, $P_{fa} = 0.1$ and $L = 9$ for different values of N.

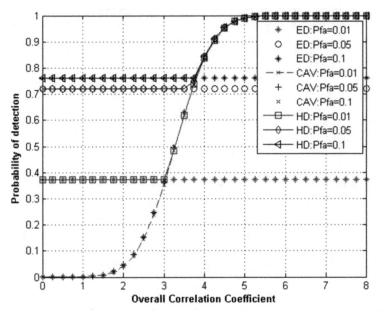

Figure 2.28 Analysis of hybrid detection, ED, and CAV methods with SNR = –20 dB, $N = 40,000$, and $L = 9$ for different values of P_{fa}.

Also the plots of P_d vs. SNR with i) $N = 40,000$, $P_{fa} = 0.01$, $\beta_L = 2$, and $L = 6$; and ii) $N = 40,000$, $P_{fa} = 0.01$, $\beta_L = 4$, and $L = 8$ are as shown in Figures 2.29 and 2.30, respectively. The performance of ED remains unchanged, and CAV becomes better for high value of correlation strength as can be easily observed by crossover points.

The performance of hybrid detection method is analyzed further by plotting P_d vs. β_L with $N = 40,000$, $P_{fa} = 0.1$, and SNR $= -20$ dB for different value of smoothing factor (L) as shown in Figure 2.31. It clearly indicates that the performance of hybrid detection algorithm becomes better if L changes from 9 to 7. Practically, it is difficult to select the best value of smoothing factor as the received signal properties are always unknown.

2.8.4 Simulation Results for Hybrid Detection Method over Fading Channels

In the following, the simulation results of hybrid detection method [77, 78], energy detection, and covariance-based detection has been evaluated for the same set of simulation parameters as shown in previous Section 2.7.3. The

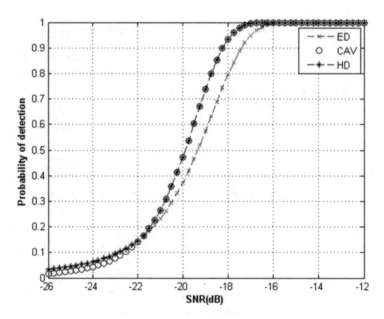

Figure 2.29   Analysis of hybrid detection, ED, and CAV with $N = 40,000$, $P_{fa} = 0.01$, $L = 6$, and $\beta_L = 2$ for different values of SNR.

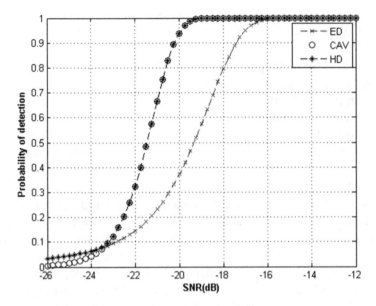

Figure 2.30 Analysis of hybrid detection, ED, and CAV with $N = 40,000$, $P_{fa} = 0.01$, $L = 8$, and $\beta_L = 4$ for different values of SNR.

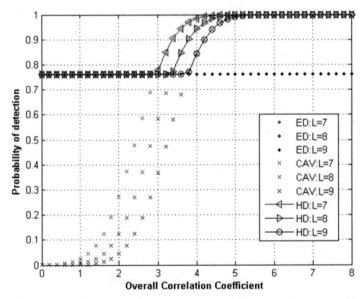

Figure 2.31 Analysis of HD, ED, and CAV with $N = 40,000$, $P_{fa} = 0.1$, and SNR $= -20$ dB for different values of smoothing factor (L).

number of samples used is 40,000. The threshold for detection in case of ED and CAV is calculated based on P_{fa} and for hybrid detection method as specified in Section 2.8.2.3.

Figure 2.32 shows the performance of ED, CAV, and HD in AWGN channel under different values of SNRs. Since ED does not depend on the overall correlation coefficient, there is a line parallel to x-axis. At SNR = –18 dB and –20 dB, ED gives probability of detection as 0.45 and 0.33, respectively, which is better than CAV at low value of correlation coefficient. For CAV, depending on overall correlation coefficient, the probability of detection varies, which approaches to 1 at β_L = 6.5 and 8 for SNR = –18 dB and SNR = –20 dB, respectively.

It is noted that hybrid detection method is performing better in low correlation compared to CAV and ED in case of high correlation. The behavior near β_{HD} can be clarified as follows: Depending on the incoming signal correlation, sometimes ED is chosen and sometimes CAV is chosen based on $\beta_L \leq \beta_{HD}$ and $\beta_L \geq \beta_{HD}$, respectively. The same is true for SNR = –20 dB.

Figure 2.33 indicates probability of detection for ED, CAV, and HD as a function of overall correlation coefficient for different values of P_{fa} with SNR = –20 dB, L = 9, and N = 40,000 in AWGN channel. ED gives

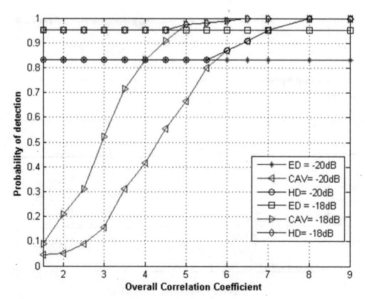

Figure 2.32 P_d for ED, CAV, and HD vs. overall correlation coefficient for different values of SNRs with P_{fa} = 0.1, L = 9, and N = 40,000 in AWGN channel.

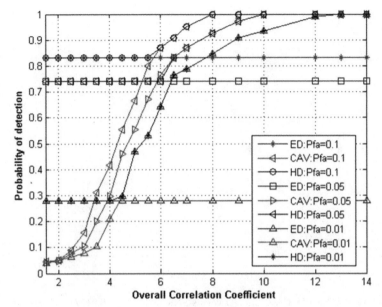

Figure 2.33 P_d for ED, CAV, and HD vs. overall correlation coefficient for different values of P_{fa} with SNR = –20 dB, $L = 9$, and $N = 40,000$ in AWGN channel.

probability of detection for P_{fa} = 0.1, 0.05, and 0.01 as 0.33, 0.24, and 0.28, respectively. For CAV depending on overall correlation coefficient, the probability of detection varies, which approaches to 1 at $\beta_L = 8, 10$, and 13 for $P_{fa} = 0.1, 0.05$, and 0.01, respectively. It is noted that hybrid detection method is performing better in low correlation compared to CAV and ED in case of high correlation.

To check the performance of the methods for time-varying channels, the time-varying channel is generated based on the simplified Jakes' model with Doppler frequency DF1 = 100 Hz and DF2 = 1000 Hz for Rayleigh channel with P_{fa} = 0.01, $L = 6$, $\beta_L = 2.5$, and $N = 40,000$. Figures 2.34 and 2.35 reflect the fact that the performance of ED and CAV degrades in fast time-varying compared to slow time-varying fading channels. The labels ED, CAV, and HD in figure are AWGN channel response values.

The hybrid detection method takes the benefit of both the methods and performs better even in fast time-varying Rayleigh channel compared to ED and CAV methods. For Rician channel, Doppler frequency of 100 Hz with P_{fa} = 0.01, $L = 6$, $\beta_L = 2.5$, and $N = 40,000$ is considered. Figure 2.36 reflects its effect on the probability of detection for ED and CAV for $K = 1$

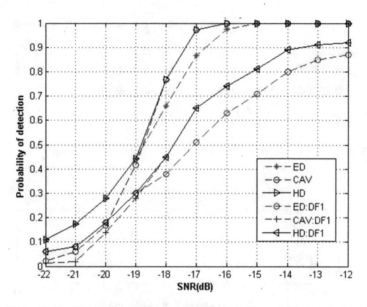

Figure 2.34 P_d for ED, CAV, and HD versus SNR with $P_{fa} = 0.01$, $L = 6$, $\beta_L = 2.5$, and $N = 40,000$ under Rayleigh time-varying channel with DF1.

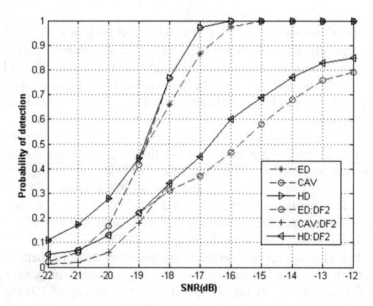

Figure 2.35 P_d for ED, CAV, and HD versus SNR with $P_{fa} = 0.01$, $L = 6$, $\beta_L = 2.5$, and $N = 40,000$ under Rayleigh time-varying channel with DF2.

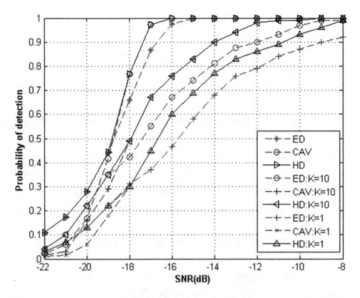

Figure 2.36 P_d for ED, CAV, and HD versus SNR with $P_{fa} = 0.01$, $L = 6$, $\beta_L = 2.5$, and $N = 40,000$ under Rician time-varying channel with $K = 10$ and $K = 1$.

and $K = 10$. Under time-varying Rician channel, the performance of both the methods degrades compared to their performance in AWGN channel.

The hybrid detection method exploits the advantage of both the methods and outperforms the two original methods. The probability of detection for ED, CAV, and HD methods in case of Rician time-varying channel with $K = 10$ is good compared to $K = 1$. The hybrid detection method becomes less sensitive to the signal correlation by taking the advantage of the both methods: by choosing energy detection in low correlation and covariance-based detection in high correlation. Furthermore, in time-varying fading channels such as Rayleigh and Rician, hybrid detection method outperforms the both original methods: energy detection and covariance-based detection. Hence, hybrid detection method can be applied to IEEE 302.11af standard for spectrum sensing in CR network exploiting TVWS.

2.9 Conclusions

For efficient utilization of spectrum opportunities for foresaid applications, spectrum sensing algorithms based on energy detection without noise uncertainty and covariance-based detection are analyzed. The simulation results show that energy detection can be applied for signal detection with a priori noise power knowledge and covariance method can be applied for signal detection application without knowledge of signal, channel and noise power. The simulation based on DVB-T signals have been carried out to evaluate the performance of both methods. The performance of energy detection and covariance-based spectrum sensing degrades in fast time-varying fading channels. This degradation in performance occurs less in case of slow time-varying channels. The energy detection without noise uncertainty outperformed the covariance-based method in case of fading time-varying channels. The probability of detection for covariance-based detection improves as smoothing factor and overall correlation coefficient increase. ED with 0.4 dB noise uncertainty shows poor probability of detection.

Further, energy detection with noise uncertainty and covariance-based detection are analyzed for DVB-T signals. It is shown that covariance-based detection outperformed energy detection with noise uncertainty present in case of fast and slow time-varying channels. The performance of covariance-based detection degrades in fast time-varying compared to slow time-varying fading channels. The probability of detection for covariance-based detection improves as smoothing factor and overall correlation coefficient increase.

Energy detection performs good in high SNR value and is not dependent on correlation of incoming signal, but suffers from noise uncertainty problem. Covariance-based detection outperforms in high-correlation environment. But in case of poor correlation, even though SNR conditions are high, its performance is poor. The proposed "hybrid detection" method takes the advantage of both the methods and it is less sensitive to the signal correlation. The simulation based on DVB-T signals has been carried out to evaluate performance of energy detection, covariance-based detection, and hybrid detection methods. The hybrid detection method becomes less sensitive to the signal correlation by taking the advantage of the both methods: by choosing energy detection in low correlation and covariance-based detection in high correlation. Furthermore, in time-varying fading channels such as Rayleigh and Rician, hybrid detection method outperforms the both original methods: energy detection and covariance-based detection. The simulation results clearly indicate that hybrid detection method can be applied to upcoming IEEE 802.11af standard based on CR networks exploiting TVWS.

References

[1] International Telecom Union. http://www.itu.int.

[2] Hossain, Ekram, Dusit Niya, and Zhu Han. *Dynamic Spectrum Access and Management in CR Networks* (Cambridge: Cambridge University Press, 2009, ISBN-13 978-0-511-58032-1).

[3] Roberson, Dennis A., Cynthia S. Hood, Joseph L. LoCicero, and John T. MacDonald. "Spectral Occupancy and Interference Studies in support of CR Technology Deployment." *Networking Technologies for Software Defined Radio Networks*, 2006. DOI 10.1109/SDR.2006.4286323.

[4] Wireless World Research Forum CR and Management of Spectrum and Radio Resources, Working Group 6 White Paper, 2 Mar 2007.

[5] FCC. "First report and order." in FCC Docket No. 02-48, Feb. 2002.

[6] Mitola, J. "Cognitive Radio: An Integrated Agent Architecture for Software Defined Radio." Ph.D. Dissertation, KTH Royal Institute of Technolology, Stockholm, Sweden, 2000.

[7] Mitola, J., and G.Q. Maguire. "Cognitive Radio: Making Software Radios more Personal." *IEEE Personal Communication* 6, no. 4 (1999): 13–18.

[8] Haykin, S. "CR: Brain-Empowered Wireless Communications." *IEEE Journal on Selected Areas in Communications* 23, no. 2 (2005).

[9] www.fcc.gov/

[10] Analysis and Design of Cognitive Radio Networks Using Game Theory: http://omidi.iut.ac.ir/SDR/2007/WebPages/07_GameTheory/CR.html [IEEEUSA_03]

[11] Chen, K-C., and R. Prasad. *Cognitive Radio Networks*. (New York: John Wiley & Sons Ltd., 2009).

[12] CR Working Group.

[13] www.sdrforum.org.

[14] Akyildiz, Ian F., W.Y. Lee, M.C. Varun, and S. Mohanty. "A Survey on Spectrum Management in CR Networks." *IEEE Communication Magazine*, April 2008.

[15] Sridhara, K., Ashok Chandra, and P.S.M. Tripathi. "Spectrum Challenges and Solutions by Cognitive Radio: An Overview." *Wireless Personal Communication* 45 (2008): 281–291.

[16] Fullmer, C.L., J. Garcia-Luna-Aceves. "Solutions to hidden terminal problems in wireless networks." *Proceedings of the ACM SIGCOMM97 Conference on Applications, Technologies, Architectures, and Protocols for Computer Communication* 97 (September 1997): pp. 39–49.

[17] Ganesan, G., and L. Ye. "Cooperative Spectrum Sensing in Cognitive Radio, Part I: Two user networks." *IEEE Transaction on Wireless Communication* 6, no. 6 (2007): 2204–2213.

[18] Peh, E., and Y.-C. Liang. "Optimization for Cooperative Sensing in Cognitive Radio networks." in *Proceedings of IEEE Wireless Communication Network Conference (WCNC)*, Hong Kong, pp. 27–32, March 2007.

[19] Ghasemi, Amir, and Elvino S. Sousa. "Spectrum Sensing in Cognitive Radio Networks: Requirements, Challenges and Design Trade-offs." CR Communications and Networks. *IEEE Communications Magazine*, pp. 32–39, April 2008.

[20] Akyildiz, Ian F., Brandon F. Lo, and Ravikumar Balakrishnan. "Cooperative Spectrum Sensing in CR Networks: A Survey." *The Journal of Physical Communication* 4 (2011): 40–62 (Available on line 2010).

[21] Sahai, A., and D. Cabric. "Spectrum Sensing: Fundamental Limits and Practical Challenges." *Proceedings of the IEEE International Symposium on New Frontiers in Dynamic Spectrum Access Networks (DySPAN)*, Baltimore, MD, November 2005.

[22] Tandra, R., and A. Sahai. "Fundamental Limits on Detection in Low SNR Under Noise Uncertainty." *Proceedings of Wireless Communication,* pp. 464–469, June 2005.

[23] Raychaudhuri, D., N.B. Mandayam, J.B. Evans, B.J. Ewy, S. Seshan, and P. Steenkiste. "CogNet-An Architectural Foundation for Experimental Cognitive Radio Networks within the Future Internet." MobiArch'06, San Francisco, CA, USA. December 2006.

[24] IMEC Research Group. "Cross-Layer Performance-Energy Modeling and Optimization for Wireless Multimedia Systems." *Scientific Report* 2006.

[25] Scoville, M., S. Berger, R.C. Reinhart, and J.E. Smith. "The Software-Defined Radio and Cognitive Radio Inter-Consortia Affiliation." *Military Communications Conference (MILCOM)*, Washington, USA, March 2006.

[26] Yucek T., and H. Arslan. "A Survey of Spectrum Sensing Algorithms for Cognitive Radio Applications." *IEEE Communications Surveys and Tutorial* 11, no. 1 (2009): 116-1.

[27] Vardoulias, G., J. Faroughi-Esfahani, G. Clemo, and R. Haines. "Blind Radio Access Technology Discovery and Monitoring for Software Defined Radio Communication Systems: Problems and Techniques." *Proceedings Internationl Conference on 3G Mobile Communication Technologies*, London, UK, pp. 306–310, March 2001.

[28] Shankar, S., C. Cordeiro, K. Challapali. "Spectrum Agile Radios: Utilization and Sensing Architectures." *Proceedings IEEE International Symposium on New Frontiers in Dynamic Spectrum Access Networks,* Baltimore, Maryland, USA, pp. 11–20, November 2005.

[29] Blossom, E. "GNU Radio: Tools for Exploring the Radio Frequency Spectrum." *Linux Journal* 2004, no. 122 (2004).

[30] Ettus, M. "Universal Software Radio Peripheral." [Online]. Available:www.ettus.com.

[31] McHenry, M., E. Livsics, T. Nguyen, and N. Majumdar. "XG Dynamic Spectrum Sharing Field Test Results." *Proceedings IEEE International Symposium on New Frontiers in Dynamic Spectrum Access Networks,* Dublin, Ireland, pp. 676–684, April 2007.

[32] Pratas, N., F. Meucci, D. Zrno, N. Prasad, A. Rodrigues, and R. Prasad. "Cogito test-bed-Wireless Research Evolved." *CR and Advanced Spectrum Management, 2009. CogART 2009. Second International Workshop on,* pp. 116–121, May 2009.

[33] Cabric, D., S. Mishra, and R. Brodersen. "Implementation Issues in Spectrum Sensing for CRs." *Proceedings of the Asilomar Conference on Signals, Systems and Computers* vol. 1, Pacific Grove, California, USA, pp. 772–776, November 2004.

[34] Cordeiro, C., K. Challapali, and D. Birru. "IEEE 802.22: An Introduction to the First Wireless Standard Based on CRs." *Journal of Communications* 1, no. 1 (2006).

[35] El-Saleh, A.A., M. Ismail, M.A.M. Ali, A.N.H. Alnuaimy. "Capacity Optimization for Local and Cooperative Spectrum Sensing in CR Networks." *World Academy of Science, Engineering and Technology* 50 (2009): pp. 69–75.

[36] Hu, W., D. Willkomm, M. Abusubaih, J. Gross, G. Vlantis, M. Gerla, and A. Wolisz. "Dynamic Frequency Hopping Communities for Efficient IEEE 802.22 Operation." *IEEE Communication Magazine,* 45, no. 5 (2007): 80–87.

[37] Kyperountas, S., N. Correal, and Q. Shi. "A Comparison of Fusion Rules for Cooperative Spectrum Sensing in Fading Channels." *EMS research Motorola,* pp. 1–6, 2009.

[38] Visotsky, E., S. Kuffner, and R. Peterson. "On Collaborative Detection of TV Transmissions in Support of Dynamic Spectrum Sharing." *Proceedings of the IEEE International Symposium on New Frontiers in Dynamic Spectrum Access Networks,* Baltimore, Maryland, USA, pp. 338–345, November 2005.

[39] Weiss, T., J. Hillenbrand, and F. Jondral. "A Diversity Approach for the Detection of Idle Spectral Resources in Spectrum Pooling Systems." *Proceedings of the 48th International Scientific Colloquium*, Ilmenau, Germany, pp. 37–38, September 2003.

[40] Mishra, S., A. Sahai, and R. Brodersen. "Cooperative Sensing Among Cognitive Radios." *Proceedings of the IEEE International Conference Communication* vol. 2, Istanbul, Turkey, pp. 19–24, May 2006.

[41] Wang, B., and K.J. Ray Liu. "Advances in CR Networks: A Survey." *IEEE Journal of Selected Topics in Signal Processing* 5, no. 1 (2011): 5–23.

[42] Chen, R., and J. Park. "Ensuring Trustworthy Spectrum Sensing in Cognitive Radio networks." *Proceedings of 1st IEEE Workshop Network Technol. Software Defined Radio Network*, pp. 110–119, 2006.

[43] Clancy, T.C., and N. Goergen. "Security in CR networks: Threats and mitigation." in *Proceedings of 3rd International Conference Cognitive Radio Oriented Wireless Network Communication (CrownCom)*, pp. 1–8, May 2008.

[44] Höyhtyä, M., A. Hekkala, M. Katz, and A. Mämmelä. "Spectrum Awareness: Techniques and Challenges for Active Spectrum Sensing." *Cognitive Wireless Networks*. pp. 353–372, June 2007.

[45] Wang, W. "Spectrum Sensing for Cognitive Radio." 3rd *International Symposium on Intelligent Information Technology Application Workshops*, pp. 410–412, November 2009.

[46] Brodersen, R.W., A. Wolisz, D. Cabric, S.M. Mishra, and D. Willkomm. "Corvus: a CR Approach for Usage of Virtual Unlicensed Spectrum." Berkeley Wireless Research Center (BWRC) White paper, 2004.

[47] Etkin, R., A. Parekh, and D. Tse. "Spectrum Sharing for Unlicensed Bands." *IEEE Journal on Selected Areas in Communications* vol. 25, pp. 517–528, April 2007.

[48] Arslan, H., and T. Yücek. "Spectrum Sensing for Cognitive Radio Applications." In: *CR, Software Defined Radio and Adaptive Wireless Systems* 3 (2007): 263–289.

[49] Zeng, Y.H., Y.C. Liang, A.T. Hoang, and R. Zhang. "Review on Spectrum Sensing for Cognitive Radio: Challenges and Solutions." *EURASIP Journal on Advances in Signal Processing* 2010, pp. 1–15.

[50] Proakis, J.G. *Digital Communications*, 4th ed. (New York: McGraw-Hill, 2001).

[51] Bhargavi, D., and C.R. Murthy. "Performance Comparison of Energy, Matched-Filter and Cyclostationarity-Based Spectrum Sensing." *Signal*

Processing Advances in Wireless Communications (SPAWC), 2010 IEEE 11th International Workshop, pp. 1–5, June 2010.

[52] Cabric, D., S. Mishra, and R. Brodersen. "Implementation Issues in Spectrum Sensing for CRs." *Proceedings of the Asilomar Conference On Signals, Systems and Computers*, vol. 1, Pacific Grove, California, USA, pp. 772–776, November 2004,.

[53] Lunden, J., V. Koivunen, A. Huttunen, and H.V. Poor. "Collaborative Cyclostationary Spectrum Sensing for Cognitive Radio systems." *IEEE Transactions on Signal Processing* 57, no. 11 (2009): pp. 4182–4195.

[54] Kapoor, S., S.V.R.K. Rao, and G. Singh. "Opportunistic Spectrum Sensing by Employing Matched Filter in CR Network." International Conference on *Communication Systems and network Technologies (CSNT)*, pp. 580–583, June 2011.

[55] Han, N., S.H. Shon, J.H. Chung, and J.M. Kim. "Spectral Correlation Based Signal Detection Method for Spectrum Sensing in IEEE 802.22 WRAN Systems." *Proceedings of IEEE International Conference on Advanced Communication Technology*, vol. 3, pp. 275–770, Feb. 2006.

[56] Chen, H.-S.W. Gao, and D.G. Daut. "Spectrum Sensing Using Cyclostationary Properties and Application to IEEE 802.22 WRAN." *Proceedings of IEEE GLOBECOM '07*, pp. 34–39, Nov. 2007.

[57] Qi, Y., T. Peng, W. Wenbo, and Q. Rongrong. "Cyclostationarity Based Spectrum Sensing for Wide Band CR." *WRI International Conference on Communications and Mobile Computing 09*, Vol. 1, pp. 107–111, Janeuary 2009.

[58] Li, H. "Cyclostationary Feature Based Quickest Spectrum Sensing in CR Systems." *IEEE 72nd Vehicular Technology Conference Fall (VTC 2010-Fall)*, pp. 1–5, September 2010.

[59] Josep, F.S., and W. Xiaodo. "GLRT-Based Spectrum Sensing for CR with Prior Knowledge." *IEEE Transactions on Communications*, pp. 28–217, July 2010.

[60] Bing, Z., and G. Lili. "Research of Spectrum Detection Technology in Cognitive Radio." *International Conference on Networks Security, Wireless Communications and Trusted Computing (NSWCTC '09)*, pp. 188–191, April 2009.

[61] Chandran, A., A. Karthik, R. Kumar, M.S. Subramania Siva, U.S. Iyer, R. Ramanathan, and R.C. Naidu. "Discrete Wavelet Transform Based Spectrum Sensing in Futuristic CRs." *International Conference Devices and Communications (ICDeCom)*, pp. 1–4, February 2011.

[62] Hur, Y., J. Park, K. Kim, J. Lee, K. Lim, C. Lee, H. Kim, and J. Laskar. "A CR (CR) Testbed System Employing a Wideband Multiresolution Spectrum Sensing (MRSS) Technique." *Proceedings IEEE Vehicular Technology Conference*, Montreal, Quebec French, Canada, pp. 1–5, September 2006.

[63] Urkowitz, H. "Energy Detection Of Unknown Deterministic Signals." *Proceedings of the IEEE*, 55, no. 4 (1967): 523–531.

[64] Carbic, D., A. Tkachenko, and R.W. Brodersen. "Experimental Study of Spectrum Sensing Based on Energy Detection and Network Cooperation." *Proceedings of the ACM International Workshop on Technology and Policy for Accessing Spectrum*, Baltimore, MD. August 2006.

[65] Zeng, Y.H., and Y.-C. Liang. "Spectrum-Sensing Algorithms for Cognitive Radio Based on Statistical Covariances." *IEEE Transactions on Vehicular Technology*, 58, no. 4 (2009): 314–1815.

[66] Zeng, Y.H., and Y.-C. Liang. "Covariance Based Signal Detections for Cognitive Radio." *DYSPAN07*, Ireland, pp. 202–207, April 2007.

[67] Challapali, K., S. Mangold, and Z. Zhong. "Spectrum Agile Radio: Detecting Spectrum Opportunities." *Proceedings of International Symposium on Advanced Radio Technologies*, Boulder, Colorado, USA, March 2004.

[68] Dandan, Z., and Z. Xuping. "SVM Based Spectrum Sensing in Cognitive Radio." *IEEE conference on Wireless communications, Networking and Mobile Computing*, pp. 1–4, October 2011.

[69] Khalaf, Z., A. Nafkha, J. Palicot, and M. Ghozzi. "Hybrid Architecture for Cognitive Radio equipment." *2010 Sixth Advanced International Conference on Telecommunications (AICT)*, pp. 49–51, May 2010.

[70] Sun, Y., Y. Liu; and X. Tan. "Spectrum sensing for CR based on higher–order statics." *4th International Conference on Wireless communications, Networking and Mobile Computing (WiCOM '08)*, pp.1–4, October 2008.

[71] Shellhammer, S.S., G. Chouinard, M. Muterspaugh, and M. Ghosh. *Spectrum Sensing Simulation Model*, July 2006. [Online]. Available: 'http://grouper.ieee.org/groups/802/22/'.

[72] http://www.etsi.org/DVB-terristerialstandards.

[73] Shellhammer, S.S., G. Chouinard, M. Muterspaugh, and M. Ghosh. *Spectrum Sensing Simulation Model*, July 2006. [Online]. Available: 'http://grouper.ieee.org/groups/802/22/'.

[74] Dhope, T., D. Simunic, and A. Kerner. "Analyzing the Performance of Spectrum Sensing Algorithms for IEEE 802.11af Standard in CR

Network." *Studies in Informatics and Control Journal, Romania* 21, no. 1 (2012): 93–100.

[75] Dhope, T., and D. Simunic. "Performance Analysis of Covariance Based Detection in CR." 35th *International Convention on MIPRO 2012.* Opatija, Croatia, pp. 737–742, May 2012.

[76] Dhope, T., D. Simunic and R. Prasad. "Hybrid Detection Method for Cognitive Radio." 13th *International conference on SoftCOM, SoftCOM2011*, Split-Hvar-Dubrovnik, pp. 1–5, September 2011.

[77] Dhope, T., and D. Simunic. "Spectrum Sensing Algorithm for CR Networks for Dynamic Spectrum Access for IEEE 802.11af Standard." *International Journal of Research and Reviews in Wireless Sensor Networks* 2, no. 1 (2012): 77–84.

[78] Simunic, D., and T. Dhope. "Hybrid Detection Method for Spectrum Sensing in CR." 35th *International Convention on MIPRO* 2012. Opatija, Croatia, pp. 765–770, May 2012.

3

Cooperative Spectrum Sensing

3.1 Introduction

The basic idea of CR is to reuse the spectrum whenever it finds the spectrum holes in wireless environment. The objectives of spectrum sensing are twofold: First, CR users should not cause harmful interference to PUs by either switching to an available band or limiting its interference with PUs at an acceptable level; and second, CR users should efficiently identify and exploit the spectrum holes for required throughput and quality-of-service (QoS). Thus, the detection performance in spectrum sensing is crucial to the performance of both primary and CR networks. However, detection performance in practice is often compromised with multipath fading, shadowing, receiver uncertainty, and even hidden primary problem due to PUs activity that is not spatially localized. To mitigate the impact of these issues, cooperative spectrum sensing has been shown to be an effective method to improve the probability of detection by exploiting spatial diversity by collaborating.

In the CR network, the location of CR users is randomly distributed [1]. Therefore, some users may suffer deep fading, shadowing, etc., while others not. On the other hand, some users may locate near to each other; hence, they experience the same path fading and suppose they have same average SNR. Thus, these CR users can form a "**cluster**" to make a cluster detection based on their same average SNR. The benefits we can get by adopting cluster structure in cooperative spectrum sensing are sensing performance improvement and sensing overhead reduction. The excellent survey of cluster-based sensing can be found in [1].

Generally, based on how the cooperative users forward their decision for making global decision, the cooperative sensing is categorized into distributed, centralized, and amplified and forward methods [2–7]. Centralized cooperative sensing is coordinated by the CR base station (CRBS). All CR users conduct the local spectrum sensing independently and then forward their observations to the CRBS. The CRBS makes the global decision based on decision rule. The existing decision rule can be divided into two categories: hard combination and soft combination [2–6]. The hard combination can reduce the number of reporting bits but degrades the sensing performance. In contrast, the soft combination has excellent performance, while it needs lots of overhead for reporting to the common receiver [2–6].

While cooperative gain such as improved detection performance and relaxed sensitivity requirement can be obtained, cooperative sensing can incur cooperation overhead. The overhead refers to any extra sensing time, delay, energy and operations devoted to cooperative sensing and any performance degradation caused by cooperative sensing.

At the end of this chapter, the readers can get familiarized with the concept of cooperative spectrum sensing and its types which are followed by simulations. Section 3.2 elaborates the need for cooperative sensing, and Section 3.3 elaborates the classification of cooperative sensing based on how cooperating CR users share the sensing data in the network: centralized, distributed, and relay-assisted. In Section 3.4, different combining techniques in cooperative sensing are given, and Section 3.5 describes the system model for cooperative spectrum sensing. The characterization of radio environment path is given in Section 3.6 followed by simulation results in Section 3.7.

3.2 Need for Cooperative Sensing

The SUs may suffer unlucky multipath and/or shadowing, receiver uncertainty, and even hidden primary user problem with respective to the primary transmitter. At the same time, its own transmissions may interfere with a primary receiver should it decide to transmit. To account for possible losses from multipath, shadowing, and building penetration, the secondary user must be significantly more sensitive in detecting than the primary receiver. The hidden primary user problem, channel/receiver uncertainty, and noise uncertainty are already discussed in Section 2.4 in detail. In Figure 3.1, multipath fading, receiver uncertainty, and shadowing are illustrated. As shown in the Figure 3.1, CR1, CR2, and CR6 are located inside the transmission range of primary transmitter (PU Tx), while CR5 and CR4 are outside the range. Due to the blocking of a house and multiple attenuated copies of the PU signal, CR6 experiences multipath and shadow fading such that the PU's signal may not be correctly detected. Moreover, CR5 and CR4 suffer from the receiver uncertainty problem because it is unaware of the PU's transmission and the existence of primary receiver (PU Rx).

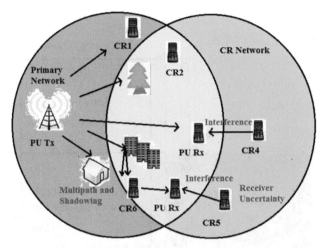

Figure 3.1   Multipath/shadow fading and receiver uncertainty.

As a result, the transmission from CR5 and CR4 may interfere with the reception at PU Rx. However, due to spatial diversity, it is unlikely for all spatially distributed CR users in a CR network to concurrently experience the fading or receiver uncertainty problem. If CR users observe a strong PU signal (in line of sight condition (LOS)) like CR1 in the Figure 3.1, they can cooperate and share the sensing results with other users. The combined cooperative decision derived from the spatially collected observations can overcome the deficiency/drawback of individual observations at each CR user. Thus, the overall detection performance can be greatly improved. This is why cooperative spectrum sensing is an attractive and effective approach to combat multipath fading and shadowing and mitigate the receiver uncertainty problem.

The main objective of cooperative sensing is to enhance the sensing performance by exploiting the spatial diversity in the observations of spatially located CR users. By cooperation, CR users can share their sensing information for making a combined decision more accurate than the individual decisions. The performance improvement due to spatial diversity is called cooperative gain [2]. The cooperative gain can be considered as the important parameter when designing the hardware for CR.

As shown in Figure 3.2, because of multipath fading, shadowing, and receiver uncertainty, SNR of the received primary signal can be extremely small and the detection of which becomes a difficult task. Since the receiver sensitivity indicates the capability of detecting weak signals, this imposes

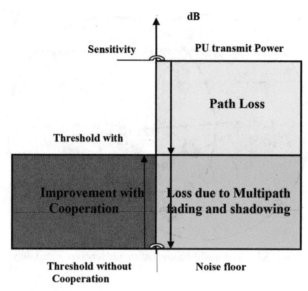

Figure 3.2   Improvement in sensitivity requirement by cooperative sensing [2].

the sensitivity constraint on the receiver greatly increasing the implementation complexity and the associated hardware cost. Sensitivity of a radio is inherently limited by cost and delay requirements. Further, the detection performance cannot be improved by increasing the sensitivity, when the SNR of PU signals is below SNR_{wall}. Degradation in the performance due to multipath fading and shadowing can be overcome by cooperative sensing such that the receiver's sensitivity can be approximately set to the same level of nominal path loss without increasing the implementation cost of CR devices [2].

By virtue of cooperative sensing, the sensitivity requirement and hardware limitation issues are relaxed. If the sensing time can be reduced due to cooperation, CR users will have more time for data transmission so as to improve their throughput. In this case, the improved throughput is also a part of cooperative gain. Thus, a well-designed cooperation mechanism for cooperative sensing can significantly contribute to a variety of achievable cooperative gain. The achievable cooperative gain can be limited by many factors. For example, when CR users blocked by the same obstacle are in spatially correlated shadowing, their observations are correlated. More spatially correlated CR users participating in cooperation can be detrimental to the detection performance. This raises the issue of user selection for cooperation in cooperative sensing.

In addition to gain-limiting factors, cooperative sensing can incur cooperation overhead. The overhead refers to any extra sensing time, delay, energy, and operations devoted to cooperative sensing compared to the individual (non-cooperative) spectrum sensing case. Moreover, any performance degradation in correlated shadowing or the vulnerability to security attacks is also a part of the cooperation overhead.

3.3 Classification of Cooperative Sensing

The cooperative sensing is divided into three categories based on how cooperating CR users share the sensing data in the network: centralized [7], distributed [8], and relay-assisted [9].

3.3.1 Centralized Cooperative Sensing

In centralized cooperative sensing, a central identity called fusion center (FC) controls the three-step process of cooperative sensing.

- First, the FC selects a channel or a frequency band of interest for sensing and instructs all cooperating CR users to individually perform local sensing.
- Second, all cooperating CR users report their sensing results via the control channel.
- Then, the FC combines the received local sensing information, determines the presence of PUs, and diffuses the decision back to cooperating CR users.

As shown in Figure 3.3, CR4 is the FC and CR0–CR3 are cooperating CR users performing local sensing and reporting the results back to CR4. For local sensing, all CR users are tuned to the selected licensed channel or frequency band where a physical point-to-point link between the PU transmitter and each cooperating CR user for observing the primary signal is called a sensing channel. For data reporting, all CR users are tuned to a control channel where a physical point-to-point link between each cooperating CR user and the FC for sending the sensing results is called a reporting channel.

Note that centralized cooperative sensing can occur in either centralized or distributed CR networks.

- In centralized CR networks, a CR base station (BS) is naturally the FC.
- Alternatively, in CR ad hoc networks (CRAHNs) where a CRBS is not present, any CR user can act as a FC to coordinate cooperative sensing and combine the sensing information from the cooperating neighbors.

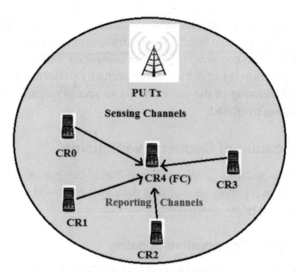

Figure 3.3   Centralized cooperative sensing.

3.3.2 Distributed Cooperative Sensing

In distributed cooperative sensing, there is no central identity (FC); that is, it does not rely on a FC for making the cooperative decision. In this case, CR users communicate among themselves and converge to a unified decision on the presence or absence of PUs by iterations.

Figure 3.4 illustrates the cooperation in the distributed manner. After performing local sensing, CR0–CR4 share the local sensing results with other users within their transmission range. Based on a distributed algorithm implemented in each CR, each CR user sends its own sensing data to other CR users, combines its data with the received sensing data, and decides whether or not the PU is present by using a local criterion. If the criterion is not satisfied, CR users send their combined results to other users again and repeat this process until the algorithm is converged and a decision is reached. In this manner, this distributed scheme may take several iterations to reach the unanimous cooperative decision.

3.3.3 Relay-Assisted Cooperative Sensing

When the reporting channels of some CRs experience heavy fading, the local decisions in these CRs cannot be forwarded to the BS. This will reduce the cooperative diversity gain. If not all CRs report to the common receiver, the

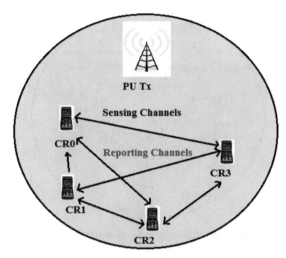

Figure 3.4 Distributed cooperative sensing.

common receiver will have to make a random decision on behalf of that CR. This will not improve the probability of detection. This can be solved by not counting unreliable reporting channels with low SNRs. However, the unreliable one can relay its local spectrum sensing result to other CRs which are in enough good channel state meaning that a CR user is observing a weak sensing channel and a strong report channel, and a CR user with a strong sensing channel and a weak report channel, for example, can complement and cooperate with each other to improve the performance of cooperative sensing. In Figure 3.5, CR0 and CR3, who observe strong PU signals, may suffer from a weak report channel. CR1 and CR2, who have a strong report channel, can serve as relays to assist in forwarding the sensing results from CR0 and CR3 to the FC. In this case, the report channels from CR1 and CR2 to the FC can also be called relay channels.

Note that although Figure 3.5 shows a centralized structure, the relay-assisted cooperative sensing can exist in distributed scheme. In fact, when the sensing results need to be forwarded by multiple hops to reach the intended receive node, all the intermediate hops behave as relays. Thus, if both centralized and distributed structures are one-hop cooperative sensing, the relay-assisted structure can be considered as multi-hop cooperative sensing. In addition, the relay for cooperative sensing here serves a different purpose from the relays in cooperative communications where the CR relays are used for forwarding the PU traffic.

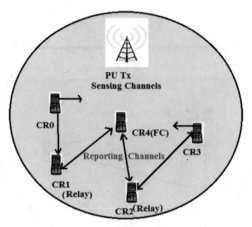

Figure 3.5   Relay-assisted cooperative sensing.

3.4 Cooperative Sensing based on Data Fusion

In cooperative sensing, data fusion is a process of combining local sensing data for hypothesis testing, which is also an element of cooperative sensing. Depending on the control channel bandwidth requirement, reported sensing results may be of different forms, types, and sizes [2]. In general, the sensing results reported to the FC or shared with neighboring users can be combined in three different ways in descending order of demanding control channel bandwidth:

- Soft combining: CR users can transmit the entire local sensing samples or the complete local test statistics for soft decision.
- Quantized soft/softened hard combining: CR users can quantize the local sensing results and send only the quantized data for soft combining to alleviate control channel communication overhead.
- Hard combining: CR users make a local decision and transmit the one-bit decision for hard combining.

Obviously, using soft combining at the FC can achieve the best detection performance among all three at the cost of control channel overhead, while the quantized soft combining/softened hard combining and hard combining require much less control channel bandwidth with possibly degraded performance due to the loss of information from quantization. In this subsection, we first discuss soft combining and quantized soft combining techniques and then focus on the fusion rules for decision fusion when the hard combining is used.

3.4.1 Soft Combining

In the soft combination scheme, sensing nodes send their sensed information directly to the decision maker without making any decisions. The decision maker makes the decision based on this received information. The soft combination gives better performance than hard combination. This is true if and only if radios are tightly synchronized in which case they can collectively overcome the SNR_{wall} (in case of energy detector). The physical noise uncertainty gives a lower bound on signal strength that a user can reliably detect. This lower bound is increased further to keep the probability of false alarm tolerable [71, 72, 2].

Existing receiver diversity techniques such as linear combining (LC) where each user computes its received source signal and sends it to the central processor, which sums the collected energy values using a linear combination (LC), equal gain combining (EGC), and maximal ratio combining (MRC), can be utilized for soft combining of local observations or test statistics [10]. In [11], an optimal soft combination scheme based on NP criterion is proposed to combine the weighted local observations. The proposed scheme reduces to EGC at high SNR and reduces to MRC at low SNR. Due to the computational complexity of the LRT-based fusion methods that involve quadratic forms, an efficient linear combination of local test statistics is proposed in [10]. In this method, the local test statistics are weighted by weighting coefficients, which are optimized based on the target P_{fa} and P_d requirements of the CR network. Since the combining weights affect the PDF of the global statistic, a modified deflection coefficient (MDC) is introduced to measure the effect of the PDF on the detector performance. Simulation results show that maximizing the MDC can result in better detection probability. This heuristic algorithm can significantly reduce the computational complexity of obtaining the global decision with a slight degradation in the detection performance. Overall, the optimal linear combination strategy is subject to performance degradation when the channel noise level increases.

3.4.2 Quantized Soft Combining

Soft combining demands a wider bandwidth for the control channel [11, 12] for receiving the sensing information data from the nodes. It also requires more overhead than the hard combination scheme. Since such a soft combining scheme results in large overhead, a softened two-bit hard combining scheme is also proposed in [11] for energy detection. The two-bit hard combination scheme [11] has the advantage of lower overhead, as demonstrated in hard combination approaches, and greater performance gain, as demonstrated in soft combination approaches. It is also called softened two-bit hard combination. The use of only one threshold in a hard combination scheme causes all nodes above the threshold to have the same weight regardless of observed energy differences between them. The main idea behind the two-bit hard combination scheme is to divide the whole range of observed energy into more than two regions and to assign different weights to these regions. By doing this, nodes that observe higher energies in upper regions have greater weights than nodes that observe lower energies in lower regions. Thus, the two-bit hard combination scheme outperforms the conventional one-bit hard combination scheme. Further, this scheme has less communication overhead when compared to the traditional soft combination schemes in which test statistics are sent to the decision maker. In this method, each node sends a two-bit information to the decision maker to inform it as to which region the observed energy fell. The three thresholds are determined by using Neyman–Pearson criterion (to meet the target overall false alarm probability of all nodes in the network) and optimizing the detection performance. A detailed threshold determination method is presented in [11]. In [13], a new three-bit hard combination scheme for collaborative spectrum sensing is proposed. Using the main idea of the two-bit hard combination scheme proposed in [11], in this case the whole range of observed energy is divided into more than four regions. In particular, seven thresholds are used to divide the whole range of observed energy into eight regions. Each node sends to the decision maker a three-bit information that indicates the region in which its observed energy fell. Dividing the range of observed energy into more than eight regions causes each node to send more than three bits of information about the observed energy region, which means more overhead. Thresholds of the three-bit hard combination scheme are determined using the Neyman–Pearson criterion. Neyman–Pearson criterion is useful when the a priori probabilities and the cost assignments for each possible decision are difficult to assign [13]. The proposed three-bit hard combination scheme is superior to the traditional hard combination schemes in false alarm reduction. The detection performance of the three bit hard combination scheme can be improved with little additional cost by increasing the number of averaged PSDs.

However, sending soft information to the FC can require much more resources than sending hard decisions. Sensor selection based on the knowledge of sensor positions is presented in [14]. In [15–17], iterative algorithms to estimate the probabilities of detection and false alarm of sensing nodes are proposed. However, these papers derive their results based on the assumption that if the local decision is the same as the final fused decision, it is the correct decision. Also in [17], weights are assigned to each CR's local sensing decision. These weights are computed by estimating the probabilities of detection and false alarm of CRs in the network. However, it is important to emphasize that the proposed methods in [15] do not select the best CRs, but only assign weights to their decisions. Moreover, it is also noteworthy that the estimation of detection and false alarm probabilities would require a large number of samples. In [18], symmetric error probabilities are assumed and a method is proposed for estimating the competence of the local decision makers without the need to know the actual value of the estimated binary event. It is important to note that the sensor selection schemes proposed in the existing literature for wireless sensor networks (WSN) are mostly formulated to guarantee the prolonged unattended deployment of the WSN for long periods of time, such as months or even years [19]. However, CRs in CR networks do not require such prolonged unattended deployment. Hence, new methods are required for intelligent selection of the CRs.

In [20], the selection of the best detection performance CRs based on only local hard decisions is studied. Three simple but efficient methods for selecting the best detection performance CRs are proposed. The best detection CRs are defined to be those CRs that have the highest probabilities of detection. The problem of selecting the CRs with the best detection performance based only on hard decisions has not been previously addressed in the context of CRs. Clearly, the smaller the set of the best detection performance CRs is used for cooperative sensing, the less the overhead and the total sensing energy is consumed because only the CRs that have the best detection performance are selected for cooperative sensing. The underlying assumption used in [14] is that a selected subset of CRs with the best detection performance will detect better in a subsequent sensing period than a subset of CRs selected purely at random.

3.4.3 Hard Combining

In this scheme, each user sends its one-bit or multiple-bit decision to a FC, which deploys a fusion rule to make the final decision. Specifically, if each user only sends one-bit decision ("1" for signal present and "0" for signal absent) and no other information is available at the central processor, some commonly adopted decision fusion rules are described [2, 13] as follows. The probability of detection, P_d, and probability of false alarm, P_{fa}, for hard combining-based cooperative spectrum sensing for OR and AND rules are as follows:

3.4.3.1 Logical-OR rule

If one of the decisions is "1," the final decision is "1." Assuming that all decisions are independent, and then, the probability of detection and probability of false alarm of the final decision Q is number of CRs are

$$C_d = 1 - \prod_{m=1}^{Q} (1 - P_{d,m}) \tag{3.1}$$

$$C_{fa} = 1 - \prod_{m=1}^{Q} (1 - P_{fa,m}) \tag{3.2}$$

3.4.3.2 Logical-AND rule

If and only if all decisions are "1," the final decision is "1." The probability of detection and probability of false alarm of the final decision are

$$C_d = \prod_{m=1}^{Q} (1 - P_{d,m}) \tag{3.3}$$

$$C_{fa} = \prod_{m=1}^{Q} (1 - P_{fa,m}) \tag{3.4}$$

3.4.3.3 "*K* out of *Q*" rule

If and only if K decisions or more are "1"s, the final decision is "1." This includes "Logical-OR (LO)" ($K = 1$), "Logical-AND (LA)" ($K = Q$), and "Majority" ($K = Q/2$) as special cases. The probability of detection and probability of false alarm are

$$P_d = \sum_{m=0}^{Q-k} \binom{n}{k+m} (1 - P_{d,m})^{n-k-m} \, x (1 - P_{d,m})^{k+m} \tag{3.5}$$

$$P_{fa} = \sum_{m=0}^{n-k} \binom{n}{k+m} (1 - P_{fa,m})^{n-k-m} \, x (1 - P_{fa,m})^{k+m} \tag{3.6}$$

where n is number of CR, $P_{d,m}$ and $P_{fa,m}$ are probability of detection and probability of false alarm of the mth CR, respectively, which can be calculated using Equations (3.3) and (3.4).

3.5 Cooperative Spectrum Sensing System Model

The cooperative spectrum sensing system model is as shown in Figure 3.6. There are one primary transmitter (PU Tx), one CRBS, and number of CR users/SU, and the location of SUs is randomly distributed. The SUs suffer different fading characteristics and have various reporting channel gain. Generally, the reporting channel between SUs and CRBS is regarded as perfect. The operation of cooperative spectrum sensing is as follows. First, all CRs conduct the local spectrum sensing independently using energy detection explained in Chapter 3 and then forward their decisions to the CRBS. CRBS applies the fusion rules OR/AND to make the global decision for declaring the availability of spectrum.

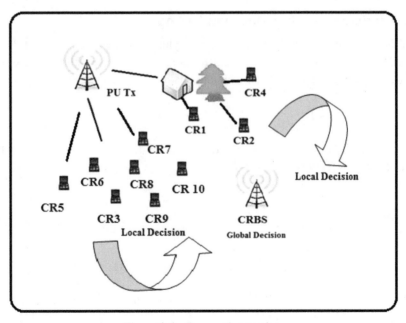

Figure 3.6 Cooperative sensing.

3.6 Characterization of the Radio Path Environment

The following scenarios for channel modeling are considered for testing the performance of cooperative spectrum sensing.

3.6.1 Outdoor Modeling, AWGN Channel

In this scenario, a line-of-sight (LOS) transmission path between transmitter and receiver is assumed to be exist, and hence, channel is modeled as AWGN with zero mean and variance σ_n^2. The probability of detection and probability of false alarm for i^{th} CR user for local energy detection sensing are given by:

$$P_{d,i} = P\{T_{ED} > \beta_{ED}|H_1\} = Q\left(\frac{\sqrt{N}\left(\beta_{ED} - \left(\sigma_s^2 + \sigma_\eta^2\right)\right)}{\sigma_s^2 + \sigma_\eta^2}\right) \qquad (3.7)$$

$$P_{fa,i} = P\{T_{ED} > \beta_{ED}|H_0\} = Q\left(\frac{\sqrt{N}(\beta_{ED} - \sigma_\eta^2)}{\sigma_\eta^2}\right) \qquad (3.8)$$

where β_{ED} is threshold for energy detector.

3.6.2 Indoor Modeling, Rayleigh Fading

In this scenario for indoor modeling, a Rayleigh fading is considered. Each ith CR user experiences independent Rayleigh fading channels with average SNR, $\bar{\varsigma}_i$; then, the PDF of the instantaneous, ς_i, is given by

$$f(\varsigma_i) = \frac{1}{\bar{\varsigma}_i} \exp\left(\frac{-\varsigma_i}{\bar{\varsigma}_i}\right), \quad \varsigma_i \ge 0 \quad i = 1,\ldots,Q \qquad (3.9)$$

The probability of false alarm and probability of detection at i^{th} CR user for local energy detection are given by

$$P_{d,i \; Rayleigh} = P\{T_{ED} > \beta_{ED}|H_1\} = e^{\frac{-\beta_{ED}}{2}} \sum_{k=0}^{\frac{N}{2}-2} \frac{1}{k!}\left(\frac{\beta_{ED}}{2}\right)^k$$

$$+ \left(\frac{1+\bar{\varsigma}_i}{\bar{\varsigma}_i}\right)^{\frac{N}{2}-1}$$

$$x\left(e^{\frac{-\beta_{ED}}{2(1+\bar{\varsigma}_i)}} - e^{\frac{-\beta_{ED}}{2}}\sum_{k=0}^{(N/2)-2}\frac{1}{k!}\left(\frac{\beta_{ED}\bar{\varsigma}_i}{2(1+\bar{\varsigma}_i)}\right)^k\right)$$

$$(3.10)$$

3.7 Simulation Results

The simulation is performed by using probability of detection as a metric at different SNR values [21]. In Figure 3.7, P_{fa} for each CR is set as 0.1, and by varying SNR from −20 to 4 dB, P_d is evaluated for OR and AND rules in AWGN channel with number of samples as 5000. We have different probability of detection which is nearly 0 for SNR value from −20 to −11 dB and reaches to 1 for OR rule at −9 dB, for AND rule at −6 dB, and for single user at −7.5 dB. These values increased as SNR improved. However, when SNR is greater than and equal to −6 dB, detection probability is relatively close and equal for both of cooperative using OR rule and non-cooperative. It means that cooperative technique using OR rule can be effectively and efficiently implemented when SNR is lower than −6 dB. In these values, cooperative technique using OR rule has better values significantly than non-cooperative one. The cooperative technique using AND rule underperforms than non-cooperative scheme.

In Figure 3.8, 2–5 collaborated users are considered for OR rule in AWGN channel. In case of low SNR, employing 5 collaborated users gives better value than the others. The low SNR is caused by propagation loss such as fading and shadowing. Comparing the single user performance in Figure 3.7 with Figure 3.8, when SNR value is greater than −7.5 dB, reveals that probability of detection obtains an optimal value relatively for both of non-cooperative and cooperative OR rule cases.

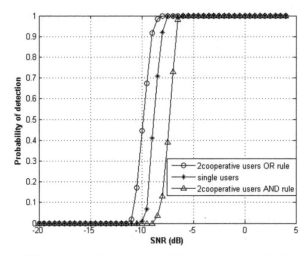

Figure 3.7 Probability of detection versus SNR in AWGN channel with N = 5000 and P_{fa} = 0.1.

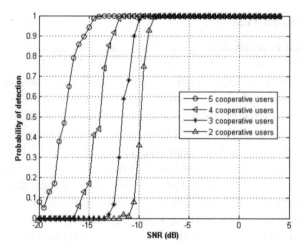

Figure 3.8 Probability of detection versus SNR for OR rule in AWGN channel with $N = 5000, P_{fa} = = 0.1$.

The evaluation of detection probability by employing AND rule in AWGN channel is shown in Figure 3.9 for 2–5 collaborated users and is compared to non-cooperative signal detection. When employing AND rule, non-cooperative case has better probability of detection values than the others. Increasing number of collaborated user causes probability of detection values

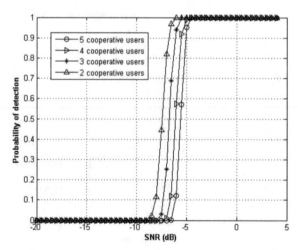

Figure 3.9 Probability of detection versus SNR for AND rule in AWGN channel with $N = 5000$ and $P_{fa} = 0.1$.

to become low in comparison with non-cooperative case. When SNR value is greater than –6 dB, probability of detection achieves an optimal and equal value relatively. Furthermore, improvement of detection probability by increasing number of collaborated user is shown in Figure 3.10. The simulation is conducted by varying number of cooperative users. We adopt OR rule and vary SNR values –10 dB, –8 dB, and –6 dB, respectively. The result shows that increasing number of collaborated users that are populated in the range of primary transmitter can improve probability of detection in CR system.

The ROC curve is plotted in Figure 3.11 for AWGN channel which reveals that there is performance improvement for 5 and 4 collaborated users compared to single user which is not able to detect PU at SNR –14 dB.

The performance of cooperative sensing in Rayleigh channel is analyzed with $P_{fa} = 0.1$ in Figures 3.12–3.16. In Figures 3.12 and 3.13, it is plotted by taking number of samples as $N = 5000$ and 10,000 for OR rule, respectively, using different collaborating users from 2 to 5. It clearly reveals that as the number of samples increases, the probability of detection in collaboration improves.

From Figure 3.12 and Figure 3.13, it clearly indicates that as the number of samples and number of collaborated users increase, the probability of detection also increases. In Figure 3.14, probability of detection versus SNR in Rayleigh channel with $N = 10,000$ and probability of false alarm 0.1 is considered for

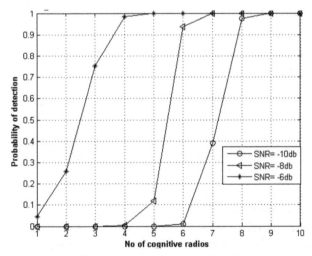

Figure 3.10 Probability of detection versus numbers CRs for OR rule in AWGN channel with $N = 5000$ and $P_{fa} = 0.1$.

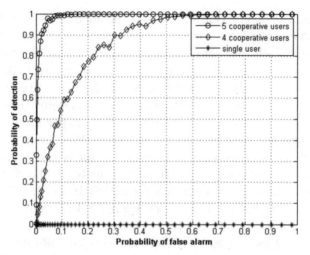

Figure 3.11   ROC for OR rule in AWGN channel with $N = 5000$ and SNR $= -14$ dB.

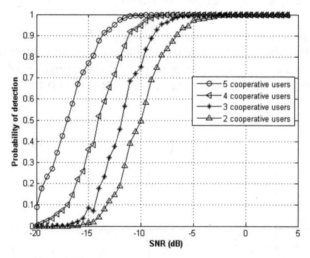

Figure 3.12   Probability of detection versus SNR for OR rule in Rayleigh channel with $N = 5000$ and $P_{fa} = 0.1$.

different collaborated users. The performance improvement is observed for OR rule compared to single user as shown in Figure 3.13. Also in Figure 3.15, the performance of cooperative sensing in Rayleigh channel is analyzed with $N = 10000$ when employing AND rule. Increasing number of collaborated user in AND rule causes probability of detection values to become low in

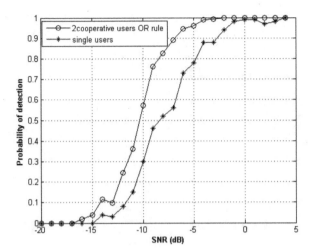

Figure 3.13 Probability of detection versus SNR in Rayleigh channel with $N = 10000$ and $P_{fa} = 0.1$.

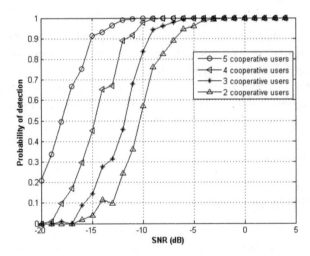

Figure 3.14 Probability of detection versus SNR for OR rule in Rayleigh channel with $N = 10,000$ and $P_{fa} = 0.1$.

comparison with non-cooperative case as shown in Figure 3.13. The non-cooperative case has better probability of detection values than the cooperative AND rule.

The numerical results show that cooperative technique has better performance compared with non-cooperative one and employing OR rule

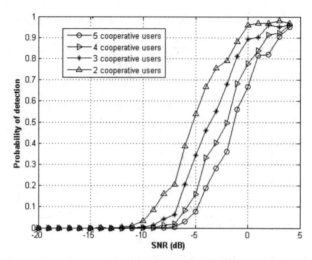

Figure 3.15   Probability of detection versus SNR for AND rule in Rayleigh channel with $N = 10,000$ and $P_{fa} = 0.1$.

can improve probability of detection than employing AND rule and non-cooperative signal detection at different SNR values. Cooperative technique is more effective when received SNR in CR users is low due to multipath fading, shadowing, and receiver uncertainty. In cooperative spectrum sensing, as the number of sample increases, probability of detection improves. Also a minimum of 10 collaborated users relatively in CR system can achieve optimal value of detection probability. However, it depends on the threshold value used in signal detection.

3.8 Conclusions

In this chapter, the performance of cooperative spectrum sensing and signal detection based on hard decision combining technique in decision fusion center compared with non-cooperative one under AWGN and Rayleigh channels has been studied. The numerical results show that cooperative technique has better performance compared with non-cooperative one and employing OR rule can improve probability of detection than employing AND rule and non-cooperative signal detection at different SNR values. Cooperative spectrum sensing technique is more effective when received SNR in CR users is low due to multipath fading, shadowing, and receiver uncertainty. In cooperative spectrum sensing, as the number of sample increases, probability of detection improves. Also a minimum of 10 collaborated users relatively in CR system can achieve optimal value of detection probability. However, it depends on the threshold value used in signal detection.

References

[1] Dhope, T., and D. Simunic. "Cluster Based Cooperative Spectrum Sensing in CR Networks: A Survey," Accepted and presented in International conference on Communication, Information and Computing Technology (ICCICT)-2012, Mumbai, India, Oct. 18th to 20th 2012.

[2] Akyildiz, I. F., B.F. Lo, Ravikumar Balakrishnan. "Co-operative Spectrum Sensing in Cognitive Radio Networks: A survey." *The Journal of Physical Communication*, 4 (2011): 40–62, (Available on line 2010).

[3] Ghasemi, Amir, and Elvino S. Sousa. "Spectrum Sensing in Cognitive Radio Networks: Requirements, Challenges and Design Trade-offs." Cognitive Radio Communications and Networks. *IEEE Communications Magazine*, pp. 32–39, April 2008.

[4] Sahai, A., and D. Cabric. "Spectrum Sensing: Fundamental Limits And Practical Challenges." In *Proceedings of the IEEE International Symposium on New Frontiers in Dynamic Spectrum Access Networks (DySPAN)*, Baltimore, MD, November 2005.

[5] Tandra, R., and A. Sahai. "Fundamental Limits on Detection in Low SNR Under Noise Uncertainty." *Proceedings of Wireless Comunication*, pp. 464–469, June 2005.

[6] Mishra, S., A. Sahai, and R. Brodersen. "Cooperative Sensing among Cognitive Radios." *White Paper*, pp. 1–7, 2005.

[7] Unnikrishnan, J., and V. Veeravalli. "Cooperative Sensing for Primary Detection in Cognitive Radio." *IEEE Journal of Selected Topics in Signal Processing* 2, no. 1 (2008): 18–27.

[8] Ganesan, G., and Y. G. Li. "Cooperative Spectrum Sensing in Cognitive Radio—Part I: Two User Networks." *IEEE Transactions on Wireless Communications*, 6, no. 6 (2007): 2204–2213.

[9] Zhang, W., and K. Letaief. "Cooperative Spectrum Sensing with Transmit and Relay Diversity in CR Networks." *IEEE Transactions on Wireless Communications* 7, no. 12 (2008): 4761–4766.

[10] Quan, Z., S. Cui, and A. Sayed. "Optimal Linear Cooperation for Spectrum Sensing in CR Networks." *IEEE Journal of Selected Topics in Signal Processing* 2, no. 1 (2008): 28–40.

[11] Ma, J., G. Zhao, and Y. Li. "Soft Combination and Detection for Cooperative Spectrum Sensing in CR Networks." *IEEE Transactions on Wireless Communications* 7, no. 11 (2008): 4502–4507.

[12] Wang, B., K. Ray Liu, and T. Clancy. "Evolutionary Cooperative Spectrum Sensing Game: How to Collaborate?" *IEEE Transactions on Communications* 58, no. 3 (2010): 890–900.

[13] Sönmezer, Volkan. "Cooperative Wideband Spectrum Sensing and Localization Using Radio Frequency Sensor Networks." Master's thesis, Naval Postgraduate School, Monterey, California, September 2009.

[14] Barkat, Mourad. Signal Detection and Estimation, 2nd ed. Boston: Artech House, 2005.

[15] Khan, Z., J. Lehtomäki, K. Umebayashi, and J. Vartiainen. "On the Selection of the Best Detection Performance Sensors for Cognitive Radio Networks." *Signal Processing Letters* 17, no. 4 (2010): 359–336.

[16] Chen, L., J. Wang, and S. Li. "An Adaptive Cooperative Spectrum Sensing Scheme Based on the Optimal Data Fusion Rule." *Proceedings of the International Symposium on Wireless Communication Systems,* 2007, pp. 582–586.

[17] Chair, Z., and P. Varshney. "Optimal Data Fusion in Multiple Sensor Detection Systems." *IEEE Transaction on Aerospace Electronic Systems* AES-22, no. 1 (1986): 98–101.

[18] Ansari, N., J. -G. Chen, and Y. -Z. Zhang. "Adaptive Decision Fusion for Unequiprobable Sources." *Proceedings of the Institute of Electrical and Electronics Engineers, Radar, Sonar and Navigation* 144, no. 3 (1997): 105–111.

[19] Mitina, O., and N. K. Vereshchagin. "How to Use Expert Advice in the Case When Actual Values of Estimated Events Remain Unknown." *COLT*, pp. 91–97.

[20] Rowaihy, H., S. Eswaran, M. Johnson, D. Verma, A. Noy, T. Brown, and T. Porta. "A Survey of Sensor Selection Schemes In Wireless Sensor Networks." *Proceedings of the Distributed Sensor Systems (DSS'07)*, Hawaii, pp. 1–13, August 2007.

[21] Dhope, T., S. S. Patil, V. Rajeshwarkar, and D. Simunic. "Performance Analysis of Hard Combing Schemes in Cooperative Spectrum Sensing for Cognitive Radio Networks." *IJEAT*, 2, no. 3 (2013): 21–26. ISSN: 2249–8958.

4

DoA Estimation Algorithms

4.1 Introduction

The high demand on the usage of the wireless communication system calls for higher system capacities. The system capacity can be improved either by enlarging its frequency bandwidth or by allocating new portion of frequency spectrum to wireless services. But since the electromagnetic spectrum is a limited resource, it is not possible to get a new spectrum allocation without the international coordination on the global level. Therefore, an efficient use of the existing spectrum is of prime interest as a research objective. Efficient source and channel coding as well as reduction in transmission power, transmission bandwidth or both are significantly contributing to this challenging issue. With the advances in digital techniques, the frequency efficiency can be improved by multiple access technique (MAT), which improves mobile users' access to the scarce resources of base station and hence improves the system's capacity [1]. By adding a new parameter of 'space' or 'angle' to the existing family of Frequency Division Multiple Access (FDMA), Time Division Multiple Access (TDMA), Code Division Multiple Access (CDMA) and Orthogonal Frequency Division Multiple Access (OFDMA) schemes, a new MAT known as 'Space Division Multiple Access' (SDMA) is established [2].

Generally, at the receiver's side, the signal received is a superposition of multipath components combined with interferers' signals, and with present noise. Thus, detection of the desired signal is a tough task. The Smart Antenna System (SAS) employs the antenna elements and the digital signal processing which enables it to form a beam to a desired direction taking into account the multipath signal components. In this way, signal-to-interference-and-noise ratio (SINR) improves by producing nulls towards the interferers in the direction of signal-of-not-interest (SoNI) [3, 4]. The performance of SAS greatly depends on the performance on direction of arrival (DoA)/angle of arrival (AoA) estimation.

113

A lot of work has been done in this area in regard to DoA estimation based on Bartlett, Capon and MUSIC in [2, 5, 6], performance analysis of MUSIC and ESPRIT in [7–10] and performance analysis of root-MUSIC in [11–13]. In [2], the normalized spatial spectrum is plotted for three users to compare the results of Bartlett, Capon and MUSIC. In [5], the relationship between the MSE and system resolution with phase mismatch is calculated for Capon and MUSIC. Here the simulation results show the dependency of MSE with the quality of array calibration. In [6], accuracy analysis in DoA estimation is done for two users based on MUSIC, Capon and root-MUSIC. In [7], the accuracy in DoA estimation is evaluated for two and three users based on MUSIC and ESPRIT.

The accuracy in DoA estimation by MUSIC, SAGE and ESPRIT is evaluated in [8] for BPSK signal. The performance of the three algorithms was compared based on their capability to estimate DoA. In [9], the performance of MUSIC and ESPRIT algorithms for narrowband signal detection and localization is analysed. The design of smart antenna system using MUSIC and ESPRIT for DoA estimation is evaluated for six-port reflectometer in [10].

In [11], the performance of the root-MUSIC is analysed by examining the perturbation in the roots of the polynomial of root-MUSIC. Further, simple closed-form expressions are derived for the one and two users to get insight. Qi Cheng [12] deals with the selection of root in root-MUSIC closest to unit circle and picking up the root which give rise to minimum value of MUSIC polynomial in order to improve the performance of root-MUSIC in blind carrier frequency offset estimation.

But most of these studies have not made attention in finding out mean-squared error (MSE) in DoA estimation, evaluating standard deviation of DoA estimation, in analysing the effect of user space distribution such as narrow angular separation, wide angular separation and combination of narrow and wide angular separation on the performance of these algorithms. In this chapter, the performance of Bartlett, Capon, MUSIC, ESPRIT and root-MUSIC is evaluated under different sets of environment such as number of array elements, user space distribution that is narrow angular separation, wide angular separation and combination of wide and narrow angular separation, number of snapshots, signal-to-noise ratio (SNR). MSE and standard deviation of DoA estimation are used for describing quality of DoA estimation algorithms. Further, limitations and sensitivity of all algorithms to the DoA angles around direction along the antenna array are analysed.

Further, in order to efficiently utilize the spectrum opportunity, the spectrum opportunity must be scanned in 'Angle' and 'code' dimensions other than the conventional dimensions such as time, frequency and space in CR context. With the recent advances in multi-antenna technologies through beamforming, multiple users (PU and/or SU) can be multiplexed into the same channel at the same time in the same geographical area. The performance of DoA estimation algorithms is analysed by SU to seek for the spectrum opportunity in 'Angle' dimension at the end of this chapter.

The research and comparison of performance was limited to the special case of continuous wave (unmodulated signal), two-dimensional model is used, and the antennas used were with omnidirectional pattern (dipole like). Also we did not take mutual coupling into account. The organization of this chapter is as follows. Section 4.2 describes the SAS and SDMA, Section 4.3 gives a description of various DoA estimation algorithms, Section 4.4 shows the simulation results for MUSIC, root-MUSIC and ESPRIT algorithms, Section 4.5 elaborates the simulation results for MUSIC, Capon and root-MUSIC algorithms. Section 4.6 gives insight of how DoA estimation algorithms can be used in CR context to seek spectrum opportunity in 'Angle' dimensions by SU. This is followed by simulation results in Section 4.7 and proposal for novel DoA estimation algorithm.

4.2 Smart Antenna System and SDMA

As shown in Figure 4.1, SDMA can be realized using SAS which is a self-organizing system. SAS can locate and track signals, dynamically adjust the antenna pattern to enhance a reception, while minimizing interference using signal processing algorithms [3, 14–16]. SAS generates multiple beam patterns; each beam would be assigned to one user, improving frequency reuse capability and increase in channel capacity. In digital beamforming antenna systems, the signals are detected and digitized at array element level. RF signals are down-converted to baseband signals of two components: I (in phase) and Q (quadrature) information [1, 4, 17].

Each antenna system performs DoA estimation to find signal-of-interest (SoI). DoAs/AoAs of all the signals are computed by calculating time delays between antenna elements. This is done in parameter estimator block. The output of parameter estimator block is fed to adaptive beamforming where DSP uses cost (error) function for calculating the optimum filter weights using adaptive algorithm (least mean square (LMS)) that generates an array factor for an optimal signal-to-interference ratio (SIR).

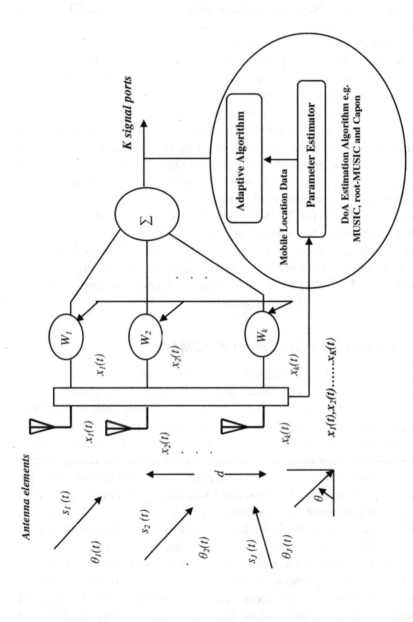

Figure 4.1 SDMA employing Smart Antenna System (θ_1 to θ_J are DoAs of transmitted signals s_1 to s_J; x_1 to x_K are antenna-specific received signals; $d = 0.5\,\lambda$).

Specifically, this results in an array pattern, where ideally the maximum of the pattern is placed towards the intended user (SoI) while nulling or attenuating the interfering users (SonI) [4, 18, 20]. Estimation of DoA entirely depends on the performance of selected DoA algorithm.

4.3 Classification of DoA

The DoA algorithms are classified as beamforming type and subspace type [2–10]. The Bartlett and Capon (Minimum Variance Distortionless Response) are beamforming type algorithms. Subspace-based DoA estimation method is based on the eigen decomposition. The observed covariance matrix is decomposed into two orthogonal spaces: signal and noise. The DoA estimation is calculated from any one of the subspaces [2–10]. The subspace-based DoA estimation algorithms MUSIC and ESPRIT provide high resolution, and they are more accurate and not limited to physical size of array aperture [9]. Various DoA estimation algorithms are evaluated as follows.

4.3.1 System Model

A uniform linear array (ULA) is considered, with J number of signals of frequncy f_0 arriving at K number of array elements which are equally spaced at distance d between the elements. The channel noise for all channels is modelled as mutually non-coherent narrowband noise at f_0. The steering vector of dimensions Kx1 corresponding to DoA at some angle θ is given by column vector:

$$v(\theta) = [e^{(-j(m-1)2\pi d \, \sin(\theta)/\lambda)}]^T \quad m = 1, 2, \ldots K \quad (4.1)$$

where $\lambda = c/f_0$ is the wavelength, c being the velocity of light and d is the spacing between antenna elements, set for this research to $0.5\,\lambda$. The columnwise combination of all J steering vectors is called array manifold matrix $\mathbf{V}$ of dimensions KxJ given by $\mathbf{V}(\theta) = [\mathbf{v}(\theta_1) : \mathbf{v}(\theta_2) : \ldots : \mathbf{v}(\theta_J)]$.

The spatial correlation (covariance) matrix for the N number of snapshots is given by:

$$\mathbf{S_x} = \frac{1}{N} \sum_{t=1}^{N} \mathbf{x}(t)\,\mathbf{x}(t)^H \quad (4.2)$$

where H denotes the Hermitian operator and $\mathbf{x}$ denotes a vector of dimensions Kx1 consisting of received signals x_k substitution of Equation (4.1) into

Equation (4.2) results in:

$$S_x = \frac{1}{N} \sum_{t=1}^{N} V(\theta) s(t) s(t)^H V(\theta)^H + n(t) n(t)^H$$

$$= V(\theta) S_s V(\theta)^H + \sigma_w^2 I \tag{4.3}$$

where σ_w^2 is noise variance, I is an identity matrix of size KxK and S_s is received signal power matrix.

4.3.2 Beamforming Techniques

The principle behind beamforming technique is to 'steer' the array in one direction at a time and measure output power. The steering locations that give maximum power yield DoA estimates. A number of sources will correspond to a number of peaks. The array response is steered by forming a linear combination of the sensor outputs [2, 5]. The array output is:

$$y = \sum_{i=1}^{K} w_i^* x_i = w^H x(t) \tag{4.4}$$

where $w = [W_1, W_2, \ldots, W_k]^T$ is a complex weighting vector, which determines the radiation pattern. The array output samples $y(1), y(2), \ldots, y(N)$ give output power as:

$$Po/p(w) = \frac{1}{N} \sum_{t=1}^{N} |y(t)|^2 = \frac{1}{N} \sum_{t=1}^{N} w^H x(t) x(t)^H w$$

$$= w^H S_x w \tag{4.5}$$

4.3.2.1 Bartlett's method

It is an extension of classical Fourier Transform–based spectrum analysis. It maximizes the power of beamforming output for a given input signal [2].

$$P_{bf} = \frac{v(\theta)^H S_X v(\theta)}{v(\theta)^H v(\theta)} \tag{4.6}$$

4.3.2.2 Capon's method

Capon's method is also called Minimum Variance Distortionless Response algorithm (MVDR). Capon's method attempts to minimize the power contributed by noise and any signals coming from other direction than desired. The

Capon's perform well with respect to Bartlett [4]. Both the methods involve in the evaluation of spectrum and then finding the local maxima which gives the estimation of DoA. The key is to minimize power contributed by noise and any signals coming from other direction than desired [2–6].

$$\min_{\mathbf{w}} (\mathbf{w}^H \mathbf{S}_x \mathbf{w}) \quad \text{Subject to } |\mathbf{w}^H \mathbf{v}(\theta)| = 1 \tag{4.7}$$

The Capon's weight vector is found to be:

$$\mathbf{w}_{\text{capon}} = \frac{\mathbf{S}_x^{-1} \mathbf{v}(\theta)}{\mathbf{v}(\theta)^H \mathbf{S}_x^{-1} \mathbf{v}(\theta)} \tag{4.8}$$

Thus, Capon's output spectrum is:

$$P_{\text{capon}} = \frac{1}{\mathbf{v}(\theta)^H \mathbf{S}_x^{-1} \mathbf{v}(\theta)} \tag{4.9}$$

4.3.3 Subspace-based Methods

In Subspace-based method, the observed covariance matrix is decomposed into two orthogonal spaces: signal subspace and noise subspace [7–10]. The DoA estimation is calculated from any one of the subspaces. The subspace-based DoA estimation algorithms MUSIC and ESPRIT provide high resolution, and they are more accurate and not limited to physical size of array aperture [7].

4.3.3.1 Music algorithm

MUSIC stands for **MU**ltiple **SI**gnal Classification, one of the high resolution subspace DoA algorithms, which gives an estimate of a number of arrived signals, hence their direction of arrival [5–9]. The estimation of DoA is performed from one of the subspaces either through signal or noise, assuming that noise in each channel is highly uncorrelated. This makes the covariance matrix diagonal. Writing the spatial covariance matrix in terms of eigenvalues and eigenvectors [4, 10, 22] gives:

$$\mathbf{S}_x = \sum_{i=1}^{K} T_i \boldsymbol{\varphi}_i \boldsymbol{\varphi}_i^H \tag{4.10}$$

$$\mathbf{S}_x = \boldsymbol{\varphi}_i \beta \boldsymbol{\varphi}_i^H \tag{4.11}$$

$$\beta = \text{diag}\left[T_1, T_2, \ldots, T_K\right] \tag{4.12}$$

The noise subspace eigenvalues and eigenvectors are:

$$T_i, i = J + 1, J + 2, \ldots, K \tag{4.13}$$

$$\varphi_i, i = J + 1, J + 2, \ldots, K \tag{4.14}$$

The noise subspaces can be written in the form of $K \times (K - J)$ matrix:

$$\vartheta_N = \left[\varphi_{J+1}, \varphi_{J+2}, \ldots, \varphi_K\right] \tag{4.15}$$

Equation (4.15) indicates that the desired value DoA of $\theta_1, \theta_2, \ldots\ldots, \theta_J$ can be found out by finding a set of vectors that span ϑ_N and projecting $\mathbf{v}(\theta)$ onto ϑ_N for all values of θ and evaluating the J values of θ, where the projection is zero:

$$\mathbf{v}_i^H \vartheta_N^2 = 0 \quad i = 1, 2, \ldots\ldots, J \tag{4.16}$$

Thus, MUSIC Pseudospectrum is given as:

$$P_{\text{music}}(\theta) = \frac{1}{\text{abs}\left[\mathbf{v}(\theta)^H \vartheta_N \vartheta_N^H \mathbf{v}(\theta)\right]} \tag{4.17}$$

4.3.3.2 Esprit algorithm

Its acronym stands for **E**stimation of **S**ignal **P**arameter via **R**otational **I**nvariance **T**echnique. This algorithm is more robust with respect to array imperfections than MUSIC. Computation complexity and storage require- ments are lower than MUSIC as it does not involve an extensive search throughout all possible steering vectors. However, it explores the rotational invariance property in the signal subspace created by two subarrays derived from original array with translation invariance structure. It is based on the array elements placed in identical displacement–forming matched pairs, with K array elements, resulting in $m = K/2$ array pairs called 'doublets' [7–10, 22]. Two subarrays P_1 and P_2 with the translation displacement $\triangle p$, finding the signals received at q^{th} doublets, are given as:

$$W_{1,q}(t) = \sum_{k=1}^{J} S_k(t)\,\alpha_q(\theta_k) + n_{1,q}(t) \tag{4.18}$$

$$W_{2,q}(t) = \sum_{k=1}^{J} S_k(t)\,\alpha_q(\theta_k) + n_{2,q}(t) \tag{4.19}$$

where θ_k is the direction of arrival of the Kth source relative to the direction of translation displacement $\triangle p$. The total output vector $u(t)$ is expressed as:

$$u(t) = \begin{bmatrix} uo(t) \\ u1(t) \end{bmatrix} = \overline{B}s(\text{t}) + \text{n(t)} \qquad (4.20)$$

Computation of the signal subspace for the two subarrays, P_1 and P_2, results in two vectors V_1 and V_2 such that Range [S] = Range [B]. In addition, there should be an existence of a non-singular matrix T of JxJ such that $V_s = \overline{B}T$, where V_s can be decomposed into V_1 and V_2:

$$V_1 = BT, \quad V_2 = B\varphi \qquad (4.21)$$

$$\varphi = diag\left[e^{jkdsin(\theta_1)}, e^{jkdsin(\theta_2)}, \ldots, e^{jkdsin(\theta_J)}\right]. \qquad (4.22)$$

If JxJ is diagonal unitary matrix with phase shifts between doublets for each DoA, there exists a unique rank J matrix FϵC such that:

$$[V_1|V_2]F = V_1W_1 + V_2W_2 = BTW_1 + B\varphi TW_1 = 0 \qquad (4.23)$$

Rearranging Equation (4.23), we get:

$$BT\psi = B\varphi \qquad (4.24)$$

where $\psi = -W_1W_2^{-1}$ and with B as full rank and sources are having distinct DoA, then:

$$\psi = -T^{-1}\varphi T \qquad (4.25)$$

If we are able to find out eigenvalues of ψ, which are diagonal elements of φ, we can estimate the DoA as:

$$a_i = e^{jkdsin(\theta_i)} \quad i = 1, 2, \ldots, J \qquad (4.26)$$

DoA can be calculated by

$$\theta_i = \sin^{-1}\left[\frac{\arg \arg (a_i)}{kd}\right] \qquad (4.27)$$

4.3.3.3 root-Music algorithm

root-MUSIC is the polynomial version of MUSIC. The array manifold matrix is expressed in polynomial form by evaluating at $z = e^{j\theta}$ [6, 11–13].

Let $C = \boldsymbol{\vartheta}_N \, \boldsymbol{\vartheta}_N^{\mathrm{H}}$ in Equation (4.17), which may be written as:

$$p_{\mathrm{music}}^{-1} = \sum_{m=1}^{K} \sum_{n=1}^{K} \exp^{(j(m-1)2\pi d \sin(\theta)/\lambda)} C_{\mathrm{mn}} A \qquad (4.28)$$

where $\mathbf{A} = \exp^{(-j(m-1)2\pi d \sin(\theta)/\lambda)}$, and C_{mn} is the entry in the mth row and n^{th} column of C. Combination of two sums into one gives Equation (4.29):

$$p_{\mathrm{music}}^{-1} = \sum_{n=1}^{K} C_q \exp^{(-j2\pi dl \sin(\theta)/\lambda)} \qquad (4.29)$$

where $C_q = \sum_{m-n=q} C_{mn}$ is the sum of the entries of C. Along the qth diagonal, polynomial representation $D\left(z\right)$ will be:

$$D\left(z\right) = \sum_{q=-K+1}^{K+1} C_q z^{-1} \qquad (4.30)$$

If the eigen decomposition corresponds to the true spectral matrix, then MUSIC spectrum $P_{\mathrm{music}}(\theta)$ becomes equivalent to the polynomial on the unit circle and peaks in the MUSIC spectrum exist as roots of polynomial lie close to the unit circle [11–13]. That is $P_{\mathrm{rmusic}}(z)|_{z=e^{j\theta}} = P_{\mathrm{music}}(\theta)$. Ideally, in the absence of noise, the poles will lie exactly on the unit circle at the locations determined by DoA. Ultimately, we calculate the polynomial and select the J roots that are inside the unit circle. A pole of polynomial, $\mathbf{D}(z)|_{z=z_q} = |z_q| = |\exp{(j \arg(z_q)|}$, will result in a peak in the MUSIC spectrum at:

$$\theta = \sin^{-1}\left\{\lambda/2\pi d\right\} \arg\left[z_q\right] \quad q = 1, 2, \ldots\ldots, J \qquad (4.31)$$

4.4 Simulation Results for MUSIC, root-Music and ESPRIT

For a reliable comparison between algorithms, 50 trials were run for each case and their results were averaged before the comparison. The MUSIC, root-MUSIC and ESPRIT techniques for DoA estimations are simulated using MATLAB. Performance of the algorithm has been analysed as a function of array elements, as a function of SNR and as a function of snapshots. The simulation has been run for four signals coming from different angles $14°, 28°, 35°, 55°$, with 500 snapshots (n), with SNR of 10 dB, and with array size of 16 (K) [22].

4.4.1 Impact of Array Elements on MUSIC Spectrum

MUSIC spectrum for varying number of array elements is shown in Figure 4.2. It indicates that as array size increases the peaks of the spectrum become sharper and hence increases the resolution capability of MUSIC.

4.4.2 Impact of SNR on MUSIC Spectrum

Figure 4.3 indicates that as SNR value decreases, peaks in spectrum start to disappear and hence decreases resolution capability of MUSIC for closely spaced signals such as 28° and 35°.

4.4.3 Impact of Snapshots on MUSIC Spectrum

Figure 4.4 indicates the ability of MUSIC to resolve closely spaced signals 28° and 35° as a function of number of snapshots. As snapshots increased from 50 to 200, resolution capability of MUSIC increases; we can clearly identify these two signals. Peaks in the spectrum become further sharper for snapshots 500, 700 and 1000.

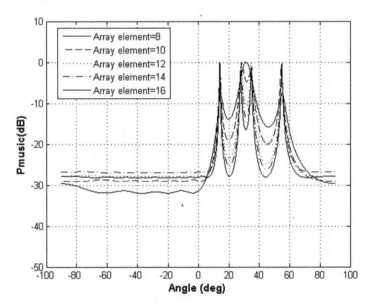

Figure 4.2 MUSIC spectrum for varying number of array elements (Note: 1 array elements = 16).

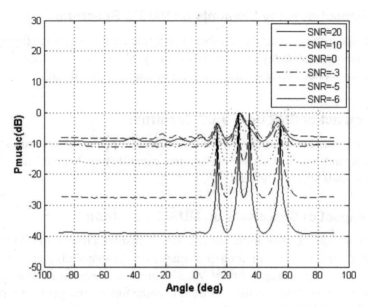

Figure 4.3   MUSIC spectrum for varying SNR (Note: 1 SNR = 5, Note: 2 SNR = 10).

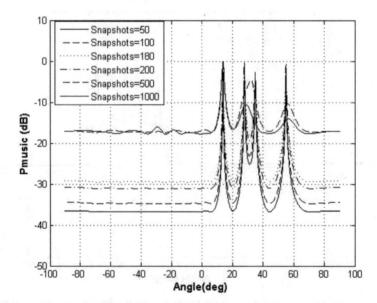

Figure 4.4   MUSIC spectrum for varying number of snapshots (Note: 1 snapshots = 500, Note: 2 snapshots = 50).

4.4.4 Impact of Snapshots (*M* = 8, SNR = 10 dB, four signals: 140, 280, 350, 550)

Table 4.1 indicates that at 200 snapshots, MUSIC cannot resolve the closely spaced signals 28° and 35°. It requires 1000 snapshots to resolve these two signals with array size of 8. ESPRIT requires 600 snapshots to resolve 28° and 35° signals accurately and root-MUSIC takes more than 2000 snapshots to resolve these two signals correctly.

4.4.5 Impact of Number of Snapshots (SNR = 10 dB, array size = 16)

Tables 4.2–4.4 indicate that for 700 and 1000 snapshots MUSIC gives an accurate estimation for four signals. If a number of snapshots increase to 1000, peaks in the spectrum become sharper and deeper as shown in Figure 4.5, resulting in the improvement of the resolution capability of MUSIC. ESPRIT and root-MUSIC identify the four signals, but the MSE is close, but not exactly zero. For snapshot value of 100, both MUSIC and ESPRIT fail to detect closely spaced signals 28° and 35°, but root-MUSIC identifies them very well: typical value is 27.13° and 35.28°, respectively. Table 4.5 reveals that MSE by MUSIC for varying number of snapshots from 500 to 1000 is zero for 14°, 35° and 55° and almost zero for 28°. For snapshots of 200 MSE by MUSIC is zero for 14° and 28°. For closely spaced signals 28° and 35° MUSIC gives MSE zero at snapshots 500, 700 and 1000 compared to other two techniques (Figures 4.5–4.7).

Tables 4.5 to 4.7 reveal that MSE for MUSIC for varying number of snapshots from 200 to 1000 is zero for 14°. For closely spaced signals 28° and 35°, MUSIC gives MSE zero at snapshots 700 compared to ESPRIT and root-MUSIC. All three algorithms give MSE zero for closely spaced signals 28° and 35° at snapshots of 1000. Figures 4.5–4.7 reveal this fact.

Table 4.1   DOA estimation for different number of snapshots with array size of 8

Snapshots	MUSIC
200	−48°, 14.6°, 27.6°, 54.1°
1000	14.2°, 27.9°, 34.6°, 54.9°
Snapshots	ESPRIT
200	13.94°, 28.90°, 37.28°, 56.42°
600	14.00°, 28.20°, 35.82°, 55.27°
Snapshots	root-MUSIC
200	−19.46°, 16.70°, 28.24°, 58.89°
2000	−20.59°, 13.97°, 30.71°, 55.28°

Table 4.2   DoA estimation by MUSIC for different number of snapshots

DoA	$n = 200$	$n = 500$	$n = 700$	$n = 1000$
14	14	14	14	14
28	28	28	28	28
35	34.9	35	35	35
55	55.8	54.9	55	55

Table 4.3   DoA estimation by ESPRIT for different number of snapshots

DoA	$n = 200$	$n = 500$	$n = 700$	$n = 1000$
14	13.94	14.01	14.99	14.03
28	28.12	27.91	27.96	28.04
35	34.77	34.98	34.94	35.01
55	55.25	55.08	54.94	54.97

Table 4.4   DoA estimation by root-MUSIC for different number of snapshots

DoA	$n = 200$	$n = 500$	$n = 700$	$n = 1000$
14	13.97	14.03	14.0	14
28	27.95	28.04	27.94	28.01
35	35.02	34.77	34.02	35.03
55	55.19	54.98	54.98	54.99

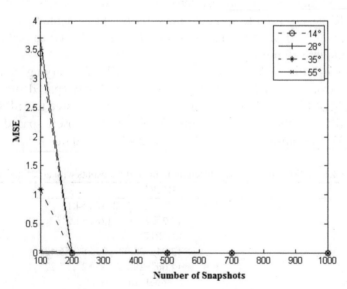

Figure 4.5   MSE by MUSIC as a function of snapshots.

Table 4.5 MSE for DoA estimation by MUSIC for different number of snapshots

DoA	$n = 200$	$n = 500$	$n = 700$	$n = 1000$
14	0	0	0	0
28	0	0.0002	0	0
35	0.0002	0.0000	0	0
55	0.0002	0.0000	0	0

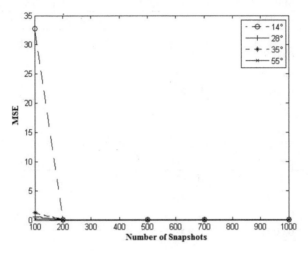

Figure 4.6 MSE by ESPRIT as a function of snapshots.

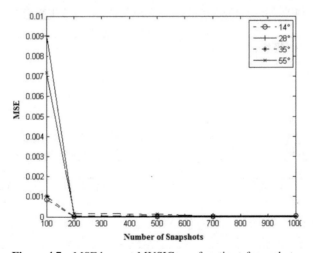

Figure 4.7 MSE by root-MUSIC as a function of snapshots.

4.4.6 Impact of Number of Array Elements (SNR = 10 dB, snapshots = 200)

Tables 4.8–4.10 indicate that the MUSIC can identify closely spaced signals at array size of 14. ESPRIT identifies 28° and 35° at array size 16. root-MUSIC also identifies at the same array size of 16 as for these values MSE is less.

Table 4.6   MSE for DoA estimation by ESPRIT for different number of snapshots

DoA	$n = 200$	$n = 500$	$n = 700$	$n = 1000$
14	0.0001	0	0	0
28	0.0003	0.0001	0	0
35	0.0010	0.0000	0.0001	0
55	0.0013	0.0001	0.0001	0

Table 4.7   MSE for DoA estimation by root-MUSIC for different number of snapshots

DoA	$n = 200$	$n = 500$	$n = 700$	$n = 1000$
14	0	0	0	0
28	0	0.0004	0.0001	0
35	0	0	0	0
55	0.0002	0.0001	0	0

Table 4.8   DoA estimation by MUSIC for different array sizes

DoA	$K = 10$	$K = 12$	$K = 14$	$K = 16$
14	13.9	14	14	14
28	28	28.2	28	28
35	35.2	34.8	35	35
55	54.9	55.1	55.1	55

Table 4.9   DoA estimation by ESPRIT for different array sizes

DoA	$K = 10$	$K = 12$	$K = 14$	$K = 16$
14	14.02	13.96	14.14	13.9
28	28.5	27.93	28.04	27.94
35	34.8	35.05	34.77	35.05
55	54.86	55.02	55.02	55.02

Table 4.10 DoA estimation by root-MUSIC for different array sizes

DoA	$K = 10$	$K = 12$	$K = 14$	$K = 16$
14	14.02	14.1	13.92	13.94
28	28.08	28.06	27.79	28.05
35	34.46	35.37	34.79	35.14
55	55.44	54.36	54.93	54.74

Tables 4.11–4.13 give idea about MSE for three algorithms. With array size 14, MUSIC gives MSE as 0. For closely spaced signals coming at 28° and 35°, MSE by MUSIC is 0.0002 and 0.0008 with array size 16. MSE by ESPRIT is 0.0001 for these both signals with array size 16. With array size 16, root-MUSIC gives MSE as 0.0001 and 0.0004 to identify DoA at 28° and 35°. Figures 4.8–4.10 clearly reveal the above fact.

4.4.7 Impact of SNR

Table 4.14 reflects the performance of algorithms for different values of SNR. As SNR decreases, the resolution capability of algorithm decreases, as well. MUSIC performs well in identifying signal even though SNR value is poor (–6 dB).

Table 4.11 MSE by MUSIC for different array sizes

DoA	$K = 10$	$K = 12$	$K = 14$	$K = 16$
14	0.0018	0.0002	0	0.0002
28	0.005	0.0008	0.0002	0.0002
35	0.0128	0.0018	0.0018	0.0008
55	0	0.0018	0.0002	0.0002

Table 4.12 MSE by ESPRIT for different array sizes

DoA	$K = 10$	$K = 12$	$K = 14$	$K = 16$
14	0	0	0.0004	0.0002
28	0.0054	0.0001	0	0.0001
35	0.0007	0.0001	0.0010	0.0001
55	0.0004	0	0	0.0003

Table 4.13   MSE by root-MUSIC for different array sizes

DoA	K = 10	K = 12	K = 14	K = 16
14	0	0.0005	0.0001	0.0001
28	0.0001	0.0001	0	0.0001
35	0.0058	0.0029	0.0008	0.0004
55	0.004	0.008	0.0001	0.0013

Table 4.14   DoA estimation by MUSIC, ESPRIT and root-MUSIC for 35°

TYPE	SNR = 10 dB	SNR = 0 dB	SNR = −6 dB	SNR = −10 dB
MUSIC	35.00	35.00	34.10	34
ESPRIT	35.11	35.74	34.99	31.56
root-MUSIC	34.97	35.26	35.32	36.6

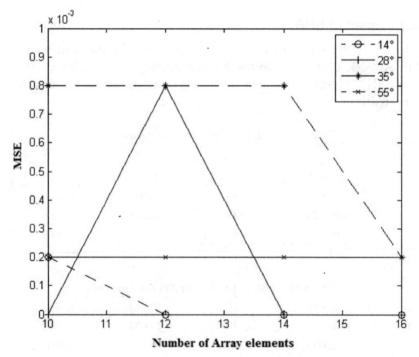

Figure 4.8   MSE by MUSIC as a function of array elements.

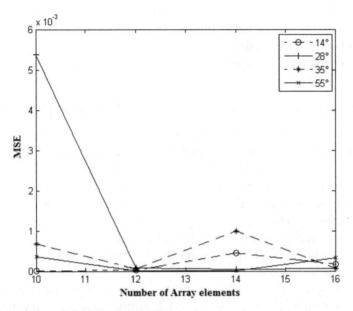

Figure 4.9 MSE by ESPRIT as a function of array elements.

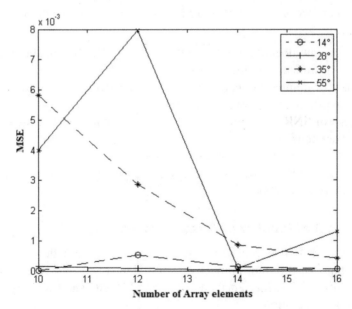

Figure 4.10 MSE by root-MUSIC as a function of array elements.

ESPRIT identifies the signal (for SNR value of –6 dB), but still error is present. root-MUSIC identifies the signal well (SNR value –6 dB) compared to ESPRIT. This reflects that if SNR decreases further, the MUSIC performs better than ESPRIT and root-MUSIC.

4.5 Simulation Results for Music, CAPON and root-MUSIC

For reliable comparison between algorithms, 50 trials were run for each case and their results were averaged before the comparison. In addition, standard deviation (in degrees) of DoA estimation (also in degrees) in these 50 trials is used for presenting accuracy and deviation of DoA estimation results: the higher the deviation, the higher the unreliability of the algorithm for given conditions. The MUSIC, Capon and root-MUSIC techniques for DoA estimations were simulated using MATLAB [23].

The simulations were run for three different sets of environment: one with wide angular separation $E_W = \{0°, 25°, 55°\}$, one with narrow angular separation $E_N = \{-5°, 10°, 20°\}$ and the last one with combination of wide and narrow separations $E_C = \{0°, 10°, 1, 40°\}$. Furthermore, accuracy in the case of DoA close to 90° (alongside the ULA) is considered [23]. For analysing the performance of these algorithms, regarding the impact of number of array elements, number of snapshots and SNR, simulation parameters were set as follows [23]:

- Impact of number of array elements: at SNR of 10 dB with 200 snapshots for environment E_W and E_N were considered.
- Impact of number of snapshots: array of 10 elements at SNR of 10 dB for environment E_N was considered.
- Impact of SNR: 16 element arrays with 200 snapshots for environment E_C were considered.
- Performance with DoAs around 90°: 16 element arrays with 200 snapshots and SNR of 10 dB were considered, with environments containing one DoA around 90°.

4.5.1 Impact of Number of Array Elements

Tables 4.15 and 4.16 as well as Figures 4.11 and 4.12 show the performance of algorithms for different number of array elements. root-MUSIC showed a significant down performance comparing to MUSIC and Capon in case of wide angular separation.

Table 4.15 Effect of number of array elements on performance of algorithms (SNR = 10 dB, snapshots = 200) in case of wide angular separation $E_W = \{0°, 25°, 55°\}$

Array Size	DoA Estimation in Degree (°) Average of 50 Trails Each for 200 Snapshots		
	MUSIC	Capon	root-MUSIC
5	0	–0.5	–60.09
	25	25.8	7.65
	55.9	56	39.29
6	0	–0.1	–28.61
	25	24.9	–0.39
	55	54.9	55.90
7	0.1	0.1	–0.09
	24.9	24.9	24.54
	55	54.9	55.99
8	0.1	0	0.14
	24.9	25	25.33
	54.8	54.8	53.66
9	0	0.1	–0.06
	24.9	25.1	24.89
	55.2	55.1	55.01
10	0	0	–0.01
	25	25	25.04
	54.9	54.9	54.84
11	–0.1	–0.1	–0.02
	25	25	25.02
	55	54.9	55.04
12	–0.1	0.1	–0.06
	25	25	25.01
	55	55	54.84
13	0	0	0.01
	25	25	25.27
	55	55	54.97
14	0	0	–0.01
	25	25	25.05
	55	55	54.99
15	0	–0.1	0.02
	25	25	24.98
	55	55.1	55.06
16	0	0	–0.02
	25	25	24.96
	55	55	54.98

Table 4.16 Effect of number of array elements on performance of algorithms (SNR = 10 dB, snapshots = 200) in case of narrow angular separation $E_N = \{-5°, 10°, 20°\}$

Array Size	DoA Estimation in Degree (°) Average of 50 Trails Each for 200 Snapshots		
	MUSIC	Capon	root-MUSIC
5	−58.3	−53.9	−34.47
	−3.9	−1.9	3.78
	10.7	12.9	69.07
6	−3.9	−38.3	−29.91
	9.9	−2.3	4.97
	19.9	13.2	46.05
7	−5	−4.1	−5.15
	9.7	12.1	15.22
	20	50	39.56
8	−4.8	−26	−5.15
	10.2	−5	10.21
	20.3	11	20.20
9	−4.9	−5.2	−4.98
	9.9	10.5	10.09
	20	19.1	20.16
10	−4.9	−5.1	−5.02
	9.9	10.2	9.98
	19.9	19.6	19.90
11	−5	−5	−4.9
	10	10.2	10.07
	20	19.8	20.13
12	−4.9	−4.9	−5.02
	10	10	9.87
	19.9	19.8	20.09
13	−5	−5	−4.99
	10	10	10.04
	20	20	19.92
14	−5	−5.1	−5.02
	10	10	9.98
	19.9	19.9	19.85
15	−5	−5	−5.08
	10	10	10.04
	19.9	20	20.01
16	−5	−5	−4.99
	10	10	10.02
	19.9	19.9	19.91

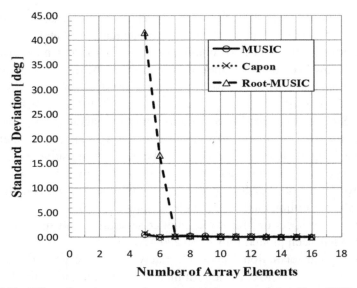

Figure 4.11 Effect of varying array elements on performance of algorithms (SNR = 10 dB, snapshots = 200) in case of wide angular separation $E_W = \{0°, 25°, 55°\}$.

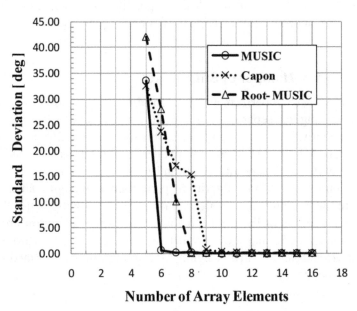

Figure 4.12 Effect of varying array elements on performance of algorithms (SNR = 10 dB, snapshots = 200) in case of narrow angular separation $E_N = \{-50, 100, 200\}$.

As shown in Figure 4.11, for wide angular separation case, MUSIC and Capon errors are less than 1° already for 5 element array and above, root-MUSIC error is still above 40° coming down to 21 for 6 element array and reaching below 1° only for 7 element array and above. As shown in Figure 4.12, for narrow angular separation case and 5 element array, all three algorithms had significant DoA estimation errors around 33°, 32°, 42° for MUSIC, Capon and root-MUSIC, respectively. MUSIC algorithm proved again to be more robust than the other two, having its error reduced to below 1° already for 6 element array case. Capon algorithm turned out to be the most sensitive to reduction in angular separation having an error of less than 1° only at 9 elements array and above, and root-MUSIC sensitivity to angular separation is higher than for MUSIC algorithm, but better than for Capon algorithm.

Further, the spectrum for wide angular separation and narrow angular separation has been plotted for Capon and MUSIC. As shown in Figure 4.13 for wide angular separation, Capon accurately detects DoA at $\{0°, 25°, 55°\}$ which is the same for MUSIC algorithm and appears in the form of sharper peaks in the spectrum for estimated DoA.

For narrow angular separation as shown in Figure 4.14, it is clearly indicated that Capon fails to identify the closely spaced signals with incidence angles of 10° and 20°. The third peak, corresponding to incidence angle of 20°, disappears in Capon spectrum, giving the poor resolution capability compared to MUSIC.

4.5.2 Impact of Number of Snapshots

Figure 4.15 shows the performance of algorithms as a function of snapshots for narrower angular separation. Below 40 snapshots, all three algorithms performed poorly. For snapshots of 10, Capon gives error as 25°, root-MUSIC nearly 23° and MUSIC 11°. This error for all three algorithms starts decreasing gradually to 10° for snapshots of 30 and to 1° at 50 snapshots. From 50 snapshots onwards, all three algorithms performed well having an error less than 1°. In this region, MUSIC algorithm performed slightly better than the other two. As shown in Table 4.17, below 50 snapshots all algorithms down perform, and above 50 snapshots onwards all three algorithms performed well having an error around 1°. In this region, MUSIC algorithm performed slightly better than the other two.

Table 4.17 Effect of varying number of snapshots on the performance of algorithm (array element = 10, SNR = 10 dB) in case of narrow angular separation $E_N = \{-5°, 10°, 20°\}$

Snapshots	DoA Estimation in Degree (°)		
	MUSIC	Capon	root-MUSIC
10	−37.4	−47.3	−36.57
	−6.5	7.5	11.41
	9.2	12.7	19.79
20	−6.6	−20.6	9.19
	9.3	−5.1	20.49
	38.9	9.5	38.74
30	−5.2	7.5	−20.73
	8.7	9.1	9.12
	36.3	37.8	20.39
40	−6.2	−6.2	−3.19
	9.3	8.8	7.78
	23.3	22.7	20.36
50	−6.1	−5.9	−4.46
	9.9	9.7	9.41
	22.4	21.7	20.01
100	−5.9	−5.7	−4.87
	9.7	9.7	10.14
	21.6	20.3	19.58
150	−5.1	−5.4	−5.01
	9.9	10	9.85
	20.3	19.8	20.16
200	−5	−5.1	−5.02
	10	10.2	9.97
	20	19.6	19.89
500	−5	−5	−5.03
	10	10	9.95
	20	20	20.02
700	−5	−5	−5.07
	10	10	9.93
	20	20	19.9

4.5.3 Impact of SNR

The performance of Capon algorithm is analysed as a function of SNR value in E_C environment by plotting its spectrum as shown in Figure 4.16. The analysis was performed using environment E_C where one separation was only 5°.

In good SNR conditions such as 10 dB and 0 dB, Capon has good resolution capability to estimate DoA at {0° 10° 15° 40°}. However, as SNR value decreases, peaks in the spectrum start to disappear and hence decreases resolution capability of capon for closely spaced signals such as 10° and 15°. From SNR of –6 dB onwards, the peak related to DoA at 1 starts disappearing.

Further, the performance of MUSIC algorithm is analysed as a function of SNR value by plotting its spectrum as shown in Figure 4.17. The analysis was performed using environment E_C where one separation was only 5°. In good SNR conditions such as 10 dB and 0 dB, MUSIC has good resolution capability to estimate DoA at {0°, 10°, 15°, 40°}. As SNR value decreases, peaks in spectrum start to disappear and hence decreases resolution capability of MUSIC for closely spaced signals such as 10° and 15°. But this effect of

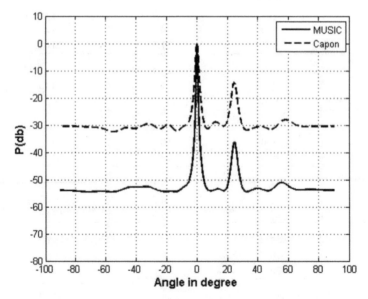

Figure 4.13 Capon and MUSIC spectrum (SNR = 10 dB, snapshots = 200, array elements = 16) and in case of wide angular separation $E_W = \{0°, 25°, 55°\}$.

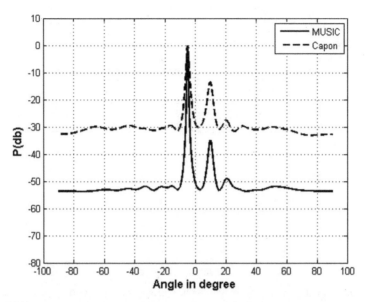

Figure 4.14 Capon and MUSIC spectrum (SNR = 10 dB, snapshots = 200, array elements = 16) in case of narrow angular separation $E_N = \{-5°, 10°, 20°\}$.

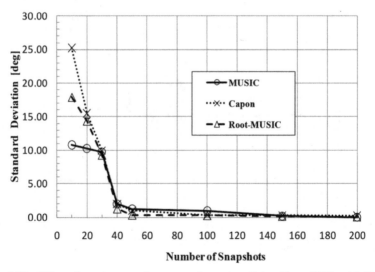

Figure 4.15 Effect of varying snapshots on performance of algorithms (SNR = 10 dB, array elements = 10) in case of narrow angular separation $E_N = \{-5°, 10°, 20°\}$.

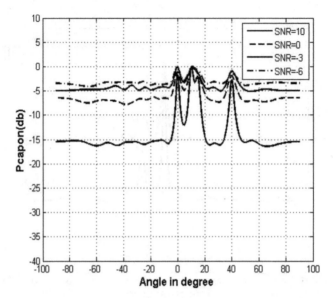

Figure 4.16   Capon Spectrum for varying SNR value (snapshots = 200, array elements = 16) in case of combination of wide and narrow angular separations $E_C = \{0°, 10°, 15°, 40°\}$.

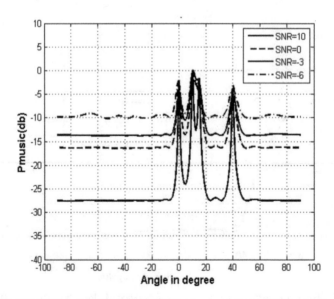

Figure 4.17   MUSIC Spectrum for varying SNR value (snapshots = 200, array elements = 16) in case of combination of wide and narrow angular separations $E_C = \{0°, 10°, 15°, 40°\}$.

vanishing peak related to DoA at 1 is not as severe compared to Capon even at SNR of –6 dB.

Figure 4.18 reveals the fact that for environment E_C, where one separation was only 5°, Capon performs poorly below 10 dB of SNR. Sensitivity of Capon towards narrower angular separation could be observed already by comparing Figure 4.12 with Figure 4.13. root-MUSIC gives DoA estimation error of 7.5° decreases to 6°, less than 1° from SNR of –15 dB to –6 dB, respectively. For lower SNRs, MUSIC again outperformed the other two algorithms.

4.5.4 Performance with DoAs around 90°

Performance of all algorithms for DoAs around 90° is analysed for case of wide angular separation (Figures 4.19, 4.20 and 4.21).

Figure 4.19 reveals that for DoA at 90°, the algorithms fail to detect it because of the appearance of false spectrum at negative 90° direction (31° flipping). Figure 4.20 reveals that for DoA at 86.5°, even though flipping is still observed in the spectrum, all the algorithms detect DoAs correctly.

Detailed analysis at what extent algorithms can detect DoAs correctly near 90° is validated in Figure 4.21 in the form of DoA detection percentage while varying DoA from 85.5° to 90°. For that purpose, simulations were run for 100 times and the percentage of correct detection was calculated.

Figure 4.21(a) reveals that all algorithms detect the DoA correctly in 100% of cases for DoAs at 85.5° and 86°. For degrees closer to 90°, there is degradation in correct DoA detection and likelihood for 31° flip in the result increases. For DoAs from 88.5° to 90°, all the algorithms fail to detect correctly, since correct detection rate is around 50%, that is as random as flipping the coin. Figure 4.21(b) reveals the all algorithms detect accurately AoAs at 85.5° and 86° (100% is the success). But for arrival angles closer to 90°, there is degradation in accurate detection of AoA and likelihood for 180° flip at the output increases. For AoAs from 89° to 90°, correct detection rate for all algorithms except MUSIC and Capon is less than 50% (as random as flipping the coin).

4.6 DoA/AoA Estimation Algorithm in Cognitive Radio Context

The conventional definition of the spectrum opportunity, which is often defined as '*a band of frequencies that are not being used by the PU of that band at a particular time in a particular geographic area*', only exploits three

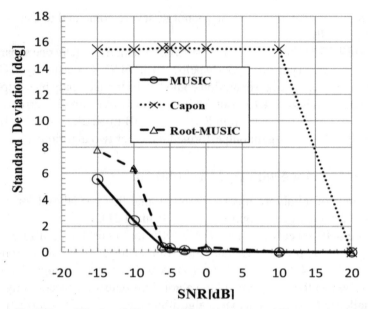

Figure 4.18 Effect of Varying SNR on algorithms (snapshots = 200, array elements = 16) in case of combination of wide and narrow angular separations $E_C = \{0°, 10°, 15°, 40°\}$.

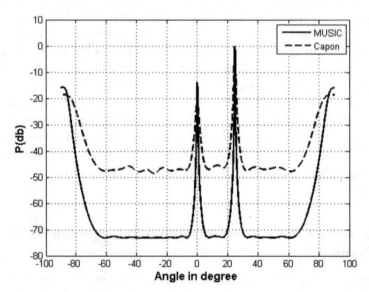

Figure 4.19 Capon and MUSIC spectrum (SNR = 10 dB, snapshots = 200, 16 element arrays) and in case of wide angular separation $E_W = \{0°, 25°, 90°\}$.

Table 4.18   Effect of varying SNR on algorithms (snapshots = 200, array elements = 16) in case of combination of wide and narrow angular separations $E_C = \{0°, 10°, 15°, 40°\}$

| | DoA Estimation in Degree (°) Average of 50 Trails | | |
| | Each for 200 Snapshots | | |
SNR(dB)	MUSIC	Capon	root-MUSIC
20	0	0	0
	10	10	9.99
	15	14.9	15
	40	40	39.98
10	0	−30.8	−0.01
	10	−0.1	10.05
	15	10.2	15.07
	40	40.1	40.01
0	0	−30.8	−0.15
	9.9	−0.2	10.08
	14.8	10.6	14.24
	40	40.1	40.18
−3	−0.1	−30.9	−0.2
	10	0.3	10.44
	14.8	10.7	14.83
	40.2	40.1	39.9
−5	−0.1	−30.9	−0.3
	10.6	−0.3	9.5
	15.1	10.7	14.1
	40.3	40	39.9
−6	−0.1	−30.9	−0.34
	10.8	−0.3	9.71
	15.3	10.7	14.16
	40.5	40	40.16
−10	−0.5	−30.9	−12.57
	10.5	−0.4	−0.42
	20	10.4	10.28
	40.8	39.8	40.55
−15	−12	−30.9	−13.75
	−8.2	−0.4	−2.9
	−0.6	10.4	11.52
	8.8	39.8	45

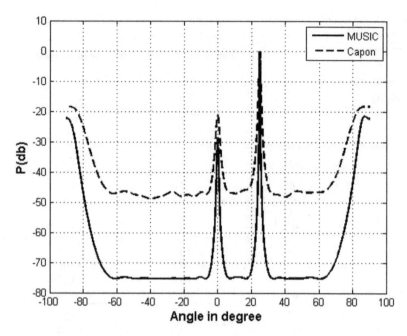

Figure 4.20   Capon and MUSIC spectrum (SNR = 10 dB, snapshots = 200, 16 element arrays) in case of narrow angular separation $E_N = \{0°, 25°, 86.5°\}$.

dimensions of the spectrum space: time, frequency and space. However, there are other dimensions 'Angle' and 'Code' that need to be explored further for spectrum opportunity [25]. The angle dimension has not been exploited well enough for spectrum opportunity. It is assumed that the PUs and/or the SUs transmit in all the directions. But with the recent advances in multi-antenna technologies through beamforming, multiple users (PU and/or SU) can be multiplexed into the same channel at the same time in the same geographical area. Here, angle dimension is different than geographical space dimension. In angle dimension, a PU and a SU can be in the same geographical area and share the same channel [25].

Geographical space dimension refers to physical separation of radios in distance. This new dimension also creates new opportunities for spectral estimation where not only the frequency spectrum but also the AoAs/DoAs need to be estimated. Once the DoA of PU signals are estimated, adaptive algorithm is used, which uses cost (error) function for calculating the optimum filter weights that maximize beam for SU towards the intended direction, a direction other than PU signal direction while nulling beam pattern in the

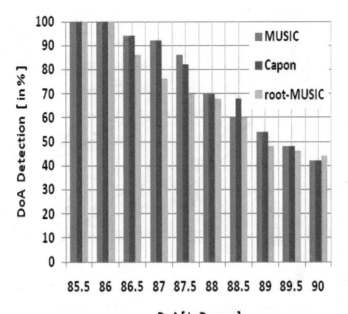

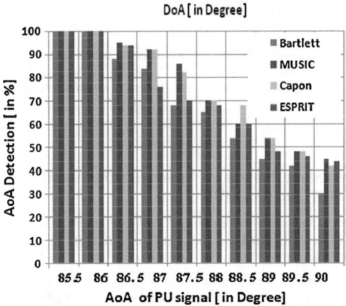

Figure 4.21 (a) DoA detection (in degree) Vs DoA from 85.5° to 90° MUSIC, Capon and root-MUSIC (SNR = 10 dB, snapshots = 200, 16 element arrays). (b) DoA detection (in degree) vs. DoA from 85.5° to 90° MUSIC, Bartlett, Capon and ESPRIT (SNR = 8 dB, snapshots = 200, 16 element arrays) [24].

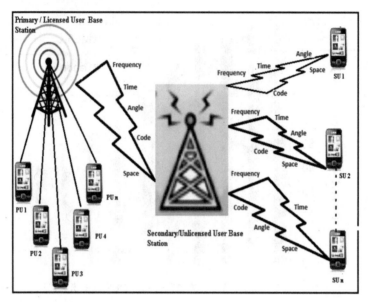

Figure 4.22   Spectrum opportunities using 'Angle' dimensions (Direction of PU beam is sensed accordingly when the direction of SU beam is decided).

direction of PU signal that enables the SU to form a beam to the direction which is not coinciding with the direction of PU signal. See Section 4.2 and Figure 4.1 in detail. Each beam can be assigned to one SU, improving frequency reuse capability and increase in channel capacity. It is also possible to multiplex multiple PUs and SUs into the same channel at the same time in the same geographical area.

4.7 Simulation Results for Capon, Bartlett's, MUSIC, ESPRIT and root-MUSIC

In order to evaluate the performance of DoA estimation algorithms for PUs in the selected frequency band, the simulations were run for three different sets of environment of PUs spatial distribution) [25]:

1. Three PUs with wide angular separation $E_W = \{0°, 25°, 55°\}$,
2. Three PUs with narrow angular separation $E_N = \{-5°, 10°, 20°\}$ and
3. Four PUs with combination of wide and narrow separations $E_C = \{0° \ 10° \ 1 \ 40°\}$.

For analysing the performance of these algorithms, regarding impact of number of array elements, snapshots and SNR parameters were set as follows:

- Impact of number of array elements were considered for two environments (E_W and E_N) at SNR of 10 dB with 200 snapshots.
- Impact of number of snapshots was considered for environment E_N with 10 array elements at SNR of 10 dB.
- Impact of SNR was considered for environment E_C with 16 array elements and 200 snapshots.

4.7.1 Impact of Array Elements

Figure 4.23 shows the performance of five algorithms for different number of array elements for $E_W = \{0°, 25°, 55°\}$ where the user spatial distribution is wide. root-MUSIC showed a significant down performance comparing to other four algorithms while MUSIC, Capon and ESPRIT errors are less than 1° for 5 element array and above. Bartlett's error is nearly 4° for 5 array elements, reaching down to less than 1° for equal and more than 6 array elements. However, root-MUSIC error is still above 25° coming down to 21 for 6 element arrays and reaching below 1° only for equal or more than 7 element arrays.

Figure 4.24 shows the performance of algorithms for different number of array elements for $E_N = \{-5°, 10°, 20°\}$. For the case of five-element array, all algorithms except ESPRIT had significant DoA estimation errors, meaning that they are either around 23° (for MUSIC, Capon and Bartlett) or higher (41° for root-MUSIC). The ESPRIT algorithm shows error less than 2°. MUSIC and ESPRIT proved to be more robust producing error below 1° for equal to or more than 6 elements array. root-MUSIC shows error below 1° for equal to or higher than 8-element array. Capon's and Bartlett's algorithms turned out to be the most sensitive to reduction in angular separation having an error of less than 1° only at equal to or more than 9- and 10-element arrays, respectively. Thus, we can conclude that for environment with narrow angular separation, root-MUSIC algorithm sensitivity to angular separation is higher than for MUSIC algorithm, but better than for Capon and Bartlett's at 8-elements array.

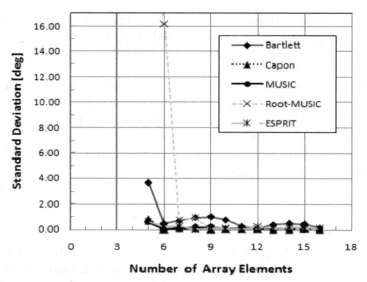

Figure 4.23 Performance analysis of DoA estimation algorithm as a function of array elements ($E_W = \{0°, 25°, 55°\}$).

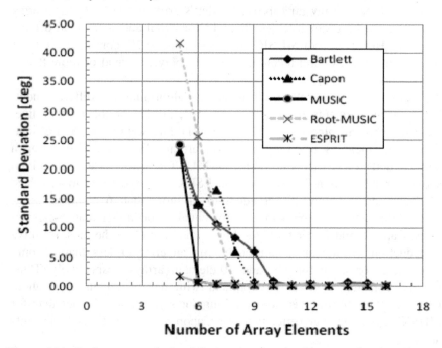

Figure 4.24 Performance analysis of DoA estimation algorithm as a function of array elements ($E_N = \{-5°, 10°, 20°\}$).

4.7.2 Impact of Snapshots

Figure 4.25 shows the performance of algorithms for different number of snapshots for $E_N = \{-5°, 10°, 20°\}$. Below 40 snapshots, all algorithms performed poorly. root-MUSIC and Capon show error more than 21 for snapshots of 10. root-MUSIC shows less than 1° errors at 50 snapshots. All algorithms perform well for 1 snapshot onwards giving highest ranking to MUSIC, ESPRIT, root-MUSIC and Capon's, respectively. Taking the excess number of snapshots results in an error of 4° in Bartlett's DoA estimation (e.g. at 700 snapshots).

4.7.3 Impact of SNR

Figure 4.26 analyses the performance of algorithm as a function of SNR with E_C where one separation was only 5°. Bartlett's and Capon's algorithm fail to identify closely spaced signals even at high value of SNR (10 dB), with error of 6° and 11°, respectively. MUSIC, root-MUSIC and ESPRIT show error below 1° at SNR of and higher to 10 dB. MUSIC outperformed the other two at SNR value of –6 dB providing error less than 1°. At SNR of –10 dB, MUSIC shows 2.5° errors, whereas root-MUSIC performs better than ESPRIT, and at SNR –5 dB, it shows error less than 1°. ESPRIT performs well only at SNRs equal to or higher than 10 dB.

4.7.4 Time-Varying Fading Rayleigh Channel

To observe the effect of time-varying fading Rayleigh channel on the performance of algorithm, the Doppler effect of 10 Hz is considered [26]. The time-varying channel is constructed using Jake's model [26].

Figure 4.27 shows the false detection in the AoA estimation under time-varying Rayleigh fading channel when the SU's velocity is assumed as 3.75 m/s at maximum carrier frequency of 802 MHz. The false detection of AoA of PU can be improved with adaptive thresholding. The adaptive thresholding sets the limit on the number of peaks detected by the algorithms so that the unwanted/undesired peaks which correspond to the undesired AoA of the signals can be filtered out. In this case, the threshold is set from 0.5 to 2 dB. The adaptive thresholding implemented for Capon and MUSIC is described here.

MUSIC and Capon without adaptive thresholding for number of samples of 500 at SNR of –6 dB for varying number of SNR from –16 to 8 dB provide more number of estimated AoA of PU signal as shown in Figures 4.28 and 4.29.

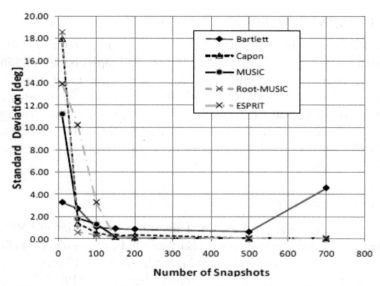

Figure 4.25 Performance analysis of DoA estimation algorithm as a function of snapshots ($E_N = \{-5°, 10°, 20°\}$).

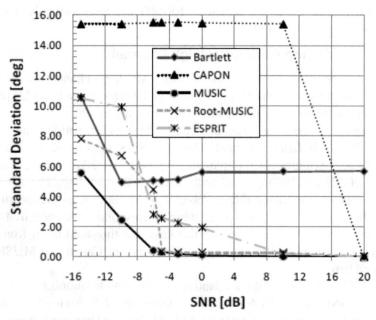

Figure 4.26 Performance analysis of DoA estimation algorithm as a function of SNR ($E_C = \{0°, 10°, 15°, 40°\}$).

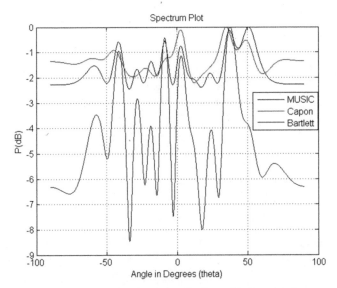

Figure 4.27 Effect of Time-varying Rayleigh Fading with Doppler frequency = 10 Hz, $W_{AS} = [-8°, 24°, 35°]$, array elements = 16, samples = 500, SNR = –6 dB.

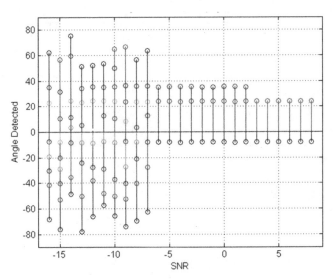

Figure 4.28 Impact of SNR for Capon without adaptive thresholding, samples = 500, array elements = 16, $E_W = [-8°\ 24°\ 35°]$.

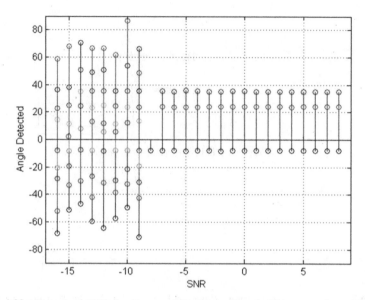

Figure 4.29   Impact of SNR for Capon with adaptive thresholding, samples = 500, array elements = 16, $E_W = [-8°\ 24°\ 35°]$.

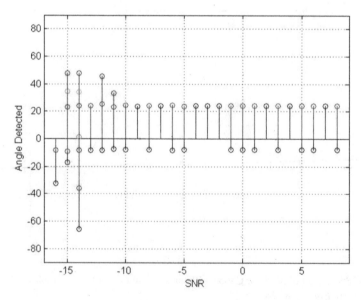

Figure 4.30   Impact of SNR for MUSIC without adaptive thresholding, samples = 500, array elements = 16, $E_W = [-8°\ 24°\ 35°]$.

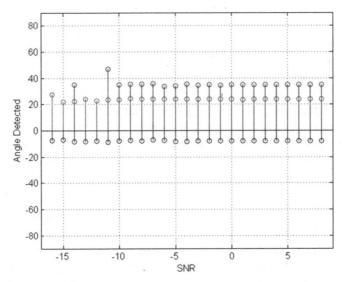

Figure 4.31   Impact of SNR for MUSIC with adaptive thresholding, samples = 500, array elements = 16, $E_W = [-8°\ 24°\ 35°]$.

Capon fails to detect 35° angle where the angular separation is only 11° for SNR values of 3 dB onwards. But with adaptive thresholding from SNR of −7 dB and above, Capon clearly identifies AoA of 35° eliminating the detection of false peaks.

MUSIC without adaptive thresholding performs better than Capon at SNR of −12 dB. It shows less number of false peaks than Capon. As shown in Figure 4.12, with adaptive thresholding, the MUSIC detects [−8° 24° 35°] AoA accurately with standard deviation of 0° from SNR of −10–8 dB.

4.7.5 Proposal of a Better Performing Algorithm

The simulation results in the previous sections of Chapter 5 clearly indicate that if one wants reliability and accuracy in DoA estimation, the recommended choice would be MUSIC algorithm, as it showed superior performance by all criteria of comparison (SNR level, number of elements, number of snapshots) and in all environment conditions (narrow, wide and combined angular separation). This new approach of DoA estimation in CR network improves frequency reuse capability and increases channel capacity by multiplexing PUs and SUs into the same channel at the same time in the same geographical area, by forming the beam of unlicensed user in the direction other than the licensed users "AoA" direction.

4.8 Conclusions

The simulation results show that performance of Capon, MUSIC, ESPRIT and root-MUSIC improves with more elements in the array, with higher number of snapshots of signals, SNR and greater angular separation between the signals. These improvements are analysed in the form of sharper peaks in MUSIC spectrum and smaller errors in angle detection. As number of snapshots increases, the MSE and standard deviation decreases which results in an accurate detection of closely spaced signals. In the analysis of impact of SNR level, at smaller levels of SNR (below 15 dB), in the combination of wide and narrow separations case with DoAs of $0°$, $10°$, $15°$ and $40°$, the performance of Capon deteriorated significantly giving standard deviation error of $15.8°$. Other two MUSIC and root-MUSIC algorithms performed well at SNRs even as low as -6 dB, and their standard deviation error was lower than 50% of Capon's in the low-SNR region below -6 dB. root-MUSIC and Capon algorithms showed to be more sensitive to small number of array elements, but Capon performed better compared to root-MUSIC. The degradation in the angle of arrival for primary user signal has been observed in time-varying fading channel. The proposed adaptive thresholding technique provides a better detection of Angle of Arrival in low SNR conditions.

If one wants reliability and accuracy in DoA estimation, the recommended choice would be MUSIC algorithm, as it showed superior performance by all criteria of comparison (SNR level, number of elements, number of snapshots) and in all environment conditions (narrow, wide and combined angular separation). These results may help when selecting tools for user's separation in cognitive radio context, in SDMA techniques and other smart antenna system designs and applications. For future fair comparisons between DoA estimation algorithms, these simulation result suggests the necessity of comparison by criteria of SNR level, number of elements and number of snapshots, but also in various environment conditions such as narrow, wide and combined angular separation.

References

[1] Litva, J., and L. Titus. *Digital Beamforming in Wireless Communication.* (London: Artech House, 1996).

[2] Shauerman, A.K., and A.A. Shauerman. "Spectral-Based Algorithms of Direction-of-Arrival Estimation for Adaptive Digital Antenna Arrays."

9th International Conference and Seminar on Micro/Nanotechnologies and Electron Devices (EDM) 2010, pp. 251–255, September 2010.

[3] Balanis, C. *Antenna Theory, Analysis and Design*. 3rd edition. (New Jersey: John Wiley and Sons, 2005).

[4] Trees, H. *Optimum Array Processing, Detection, Estimation and Modulation, Part IV*. (New York: John Wiley and Sons, 2002).

[5] Weber, R.J., and Y. Huang. "Analysis for Capon and MUSIC DoA Estimation Algorithms." *Antennas and Propagation Society International Symposium, 2009. APSURSI '09*, pp. 1–4, June 2009.

[6] Jalali, M., M. Moghaddasi, and A. Habibzadeh. "Comparing accuracy for ML, MUSIC, root-MUSIC and spatially smoothed algorithms for 2 users." *Microwave Symposium (MMS), 2009 Mediterrannean IEEE conference*, pp. 1–5, November 2009.

[7] Lavate, T.B., V.K. Kokate, and A.M. Sapkal. "Performance Analysis of MUSIC and ESPRIT DoA Estimation Algorithms for Adaptive Array Smart Antenna in Mobile Communication." 2nd International Conference on Computer and Network Technology (ICCNT), pp. 308–311, April 2010.

[8] Khan, Z.I., R.A.A Wang, A.A. Sulaiman, M.H. Jusoh, N.H. Baba, M.M.D. Kamal, and N.I. Khan. "Performance Analysis for Estimation of Signal Parameters via Rotational Invariance Technique (ESPRIT) in Estimating Direction of Arrival for linear array antenna." IEEE International Conference on RF and Microwave Conference, 2008, pp. 530–533, December 2008.

[9] Leon De, F.A., and S.J.J Marciano. "Application of MUSIC, ESPRIT and SAGE Algorithms for Narrowband Signal Detection and Localization." *TENCON, IEEE Region 10 Conference*, pp. 1–4, November 2006.

[10] Abdallah, A., S.A. Chabine, M. Rammal, G. Neveux, P. Vaudont, and M. Campovecchio. "A Smart Antenna Simulation for (DoA) Estimation Using MUSIC and ESPRIT Algorithms." 23rd *National Radio science conference (NRSC 2006)*, pp. 1–10, March 2006.

[11] Rao, B.D., and K.V.S. Hari. "Performance Analysis of Root-MUSIC." *IEEE Transactions on Acoustics, Speech and Signal Processing* 37 (1989): 1939–1949.

[12] Cheng, Qi. "Choice of Root in Root-MUSIC for OFDM Carrier Frequency Offset Estimation", *2005 Asia-Pacific Conference on Communications*, Perth, Western Australia, pp. 534–536, October 2005.

[13] Pesavento, M., A.B. Gershman, and M. Haardt. "Unitary root-MUSIC with a Real-Valued Eigen decomposition: A Theoretical and Experimental Performance Study." *IEEE Transaction on Signal Processing* 48, no. 5 (2000): 1306–1314.

[14] Kisliansky, A., and R. Shavit. "Direction of Arrival Estimation in the Presence of Noise Coupling in Antenna Arrays." *IEEE Transaction on Antennas and Propagation* 55, no. 7 (2007): 1940–1947.

[15] Blanz, J.J., A. Papathanassiou, M. Haardt, I. Furio, and P.W. Baler. "Smart Antennas for Combined DoA and Joint Channel Estimation in Time-Slotted CDMA Mobile Radio Systems with Joint Detection." *IEEE Transactions on Vehicular Technology* 49, no. 2 (2000): 293–305.

[16] Diab, W.G., and H.M. Elkamchouchi. "A Deterministic real-Time DoA-Based Smart Antenna Processor." *18th Annual IEEE International Symposium on Personal, Indoor and Mobile Radio Communications (PIMRC'07)*, pp. 1–5, September 2007.

[17] Varade, S.W., and K.D. Kulat. "Robust Algorithms for DoA Estimation and Adaptive Beamforming for Smart Antenna Application." *2nd international conference on Emerging Trends in Engineering and Technology (ICETET)*, 2009, pp. 1195–1200, December 2009.

[18] Sheng, W.X., J. Zhou, D.G. Fang, and Y.C. Gu. "Super Resolution DoA Estimation in Switched Beam Smart Antenna, Antennas." *Propagation and EM Theory, 2000. Proceedings. ISAPE 5th International Symposium*, pp. 603–606, 2000.

[19] Walid, G.D., and H.M. Elkamchouchi. "A Deterministic Real-Time DOA-based Smart Antenna Processor." Personal, Indoor and Mobile Radio Communications, 2007. PIMRC 2007. *IEEE 18th International Symposium*, 2007, pp. 1–7, 2007.

[20] Kangas, A., P. Stoica, and T. Soderstrom. "Finite Sample and Modelling Error Effects on ESPRIT and MUSIC Direction Estimators." *IEEE Proceedings-Radar, Sonar Nauig* 141, no. 5 (1994), 249–255.

[21] Akbar, M.ali, H.B. Tila, M.Z. Khalid, M.A Ajaz. "Bit Error Rate Improvement using ESPRIT based Beamforming and RAKE receiver." *IEEE 13th International Multitopic Conference, 2009. INMIC 2009*, pp. 1–6, December 2009.

[22] Dhope, T., D. Simunic, and M. Djurek. "Application of DOA Estimation Algorithms in Smart Antenna Systems." *Studies in Informatics and Control*, 19, no. 4 (2010): 445–452.

[23] Dhope, T., D. Simunic, R. Zentner. "Comparison of DoA Estimation Algorithms in SDMA System." *Automatika Journal, Croatia* 54 (2013): 199–209.

[24] Dhope, T., and D. Simunic, "On the Performance of AoA Estimation Algorithms in Cognitive Radio Networks", International conference on Communication, Information and Computing Technology (ICCICT)-2012, Mumbai, India, Oct. 18th to 20th 2012, pp. 1–5.

[25] Dhope, T., and D. Simunic, Nikhil Dhokariya, Vishal Pawar and Bhawana Gupta. "What about Spectrum Opportunities in "Angle" dimension for Dynamic Spectrum Access in Cognitive Radio context?" *2nd International Conference On Mobility for Life: Technology, Telecommunication and Problem based Learning*, 14th to 19th March 2013, pp. 198–205.

[26] Dhope, T., and D. Simunic, Nikhil Dhokariya, Vishal Pawar and Bhawana Gupta. "What about Spectrum Opportunities in "Angle" dimension for Dynamic Spectrum Access in Cognitive Radio context?" Wireless Personal Communications Special Issue: Trust and Privacy for wireless communication N/W, Springer, 75, no. 3 (2014): DOI 10.1007/s11277-014-1712-4, ISSN 0929-6212, published on line 25th Mar. 2014, pp. 1–21.

Index

Author Biography

Tanuja Shendkar Dhope, Ph.D. is a full time Associate Professor at Department of electronics and Telecommunication, G. H. Raisoni College of Engineering and Management, Pune, India. She has acquired her Ph.D in wireless communication at Faculty of Electrical Engineering and Computing, University of Zagreb, Croatia under Erasmus Mundus Mobility for Life Project in 2012 under the guidance of Dr. Dina Šimunić. She graduated in Electronics and Telecommunication engineering at Cummins College of Engineering, University of Pune in 1999. She has received Master in Electronics Engineering from Walchand college of Engineering, Sangli, Shivaji University in 2007. Her research focus is on cognitive radio network optimization with spectrum sensing algorithms, radio channel modelling for cognitive radio, wireless sensor network, cooperative spectrum sensing, Direction of arrival (DoA) Estimation algorithms in Cognitive Radio and in SDMA. She published 30 scientific papers in journals and conference proceedings. She has reviewed 15 IEEE conference papers and honoured as 'Session Chair' at ICACCI 2013 conference in Mysore.